Principles of Entrepreneurship in the Industry 4.0 Era

The feature that distinguishes this book from all other books on entrepreneurship is the comprehensive discussion of the challenges and opportunities that entrepreneurs encounter in the Industry 4.0 era. It enables readers to start dreaming big, visualizing, and encourages them to think clearly and creatively. The book emphasizes creativity and innovation as the core of entrepreneurship, by stretching imagination, thinking about problems and solutions, and visualizing their ventures at the local, national, and global scale. It also discusses the role of women in private enterprise, and entrepreneurship in the post COVID-19 world.

Features:

- Encourages possible ways to promote economic development through new venture creation.
- Helps in understanding the role of Industry 4.0 technologies in new ventures.
- Explains new concepts on how to discover ways to make start-ups more sustainable.
- Includes case studies that describe possible outcomes for various business models.
- Discusses team building as an important tool for entrepreneurial ventures.

The book will be of benefit to undergraduate and postgraduate students, Ph.D. researchers interested in entrepreneurship, and faculty teaching entrepreneurship courses in Business Administration.

Higher Education and Sustainability

J. Paulo Davim, *Professor, Department of Mechanical Engineering, University of Aveiro, Portugal*

This new series fosters information exchange and discussion on higher education for sustainability and related aspects, namely academic staff and student initiatives, campus design for sustainability, curriculum development for sustainability, global green standards: ISO 14000, green computing, green engineering education, index of sustainability, recycling and energy efficiency, strategic sustainable development, sustainability policies, sustainability reports, etc. The series will also provide information on principles, strategies, models, techniques, methodologies, and applications of higher education for sustainability. It aims to communicate the latest developments and thinking as well as the latest research activity relating to higher education, namely engineering education.

Higher Education and Sustainability
Opportunities and Challenges for Achieving Sustainable Development Goals
Edited by Ulisses Manuel de Miranda Azeiteiro and J. Paulo Davim

Designing an Innovative Pedagogy for Sustainable Development in Higher Education
Edited by Vasiliki Brinia and J. Paulo Davim

Higher Education
Progress for Management and Engineering
Edited by Carolina Machado and J. Paulo Davim

Principles of Entrepreneurship in the Industry 4.0 Era
Edited by Rajender Kumar, Rahul Sindhwani, Tavishi Tewary, and J. Paulo Davim

For more information about this series, please visit:
www.crcpress.com/Higher-Education-and-Sustainability/book-series/CRCHIGEDUSUS

Principles of Entrepreneurship in the Industry 4.0 Era

Edited by
Rajender Kumar, Rahul Sindhwani,
Tavishi Tewary, and J. Paulo Davim

CRC Press
Taylor & Francis Group
Boca Raton London New York

CRC Press is an imprint of the
Taylor & Francis Group, an **informa** business

First edition published 2023
by CRC Press
6000 Broken Sound Parkway NW, Suite 300, Boca Raton, FL 33487–2742

and by CRC Press
4 Park Square, Milton Park, Abingdon, Oxon, OX14 4RN

CRC Press is an imprint of Taylor & Francis Group, LLC

ISBN: 978-1-032-18386-2 (hbk)
ISBN: 978-1-032-18862-1 (pbk)
ISBN: 978-1-003-2-5666-3 (ebk)

DOI: 10.1201/9781003256663

Typeset in Times
by Apex CoVantage, LLC

Contents

Preface..vii
Editors..ix
Contributors xi

Chapter 1   Lean in Business Models: Way to Go for Entrepreneurial
Journey in Industry 4.0 Context 1

*Rajender Kumar, Dinesh Chawla, Sushant, Rahul Sindhwani
and Harish Kumar Banga*

Chapter 2   Mapping Entrepreneurial Resilience Using Bibliometric
Analysis—Future Agenda for Research in the Industry 4.0 Era 21

Shalini Rahul Tiwari and Shreya Homechaudhuri

Chapter 3   Trends in Social Entrepreneurship Landscape of India: Past
Contributions and Future Opportunities 39

Vikram Bansal and Deepthi B.

Chapter 4   Creating Social Capital and Outcomes through Entrepreneurship
in Industry 4.0 Perspective 53

Abhilash G. Nambudiri

Chapter 5   Study on Data-Driven Decision-Making in Entrepreneurship........... 75

Vimlesh Kumar Ojha, Sanjeev Goyal and Mahesh Chand

Chapter 6   Entrepreneurial Venture Funding and Growth in
Industry 4.0 Era... 89

Deergha Sharma and Minakshi Sehrawat

Chapter 7   Future Challenges and Opportunities in Adopting Industry
4.0 for Entrepreneurship...105

Rajeev Saha and Om Prakash Mishra

Chapter 8   Case Study on Entrepreneurial Characteristics, Attitude and
Self-Employment Intention of Undergraduates in Industry 4.0
Context 123

Pushkar Dubey, Parul Dubey and Kailash Kumar Sahu

Chapter 9   Study on Start-Ups Functioning in Industry 4.0 Context 139

Srijna Verma, Prateek Gaur, Rupali Madan and Vijay Kumar

Chapter 10   Charting Industry 4.0 Routes Incubation Centers: A Study on Atal Incubation Centre 153

Kshitiz Choudhary, Jayant Mahajan, Fr Jossy P. George and Anshul Saxena

Chapter 11   Imperatives Associated with Women's Participation in Entrepreneurial Activities 173

Om Prakash Mishra, Rajeev Saha and Rajender Kumar

Index 183

Preface

Entrepreneurship contributes to the academic, individual, and geopolitical arenas. It plays an important role in helping people survive and rise to immense stature by building private sector empires. It helps them to become independent and contribute to the process of wealth creation. Some of them even expand to international markets thereby generating mass employment and maintaining domestic stability. Against this backdrop, this book recognizes the importance of entrepreneurial activity and the role it plays in the economic development of a nation. The book encourages young scholars to pursue their dreams by stressing the need for a conducive macro-environment as a precondition for the success of an entrepreneurial journey. The book stresses the importance of individual initiatives that are required for the attainment of ecosystem promoting the culture of entrepreneurship.

The book emphasizes the fact that a relatively large number of businessmen did not amass wealth by chance or because of fortune. Specific attributes of the nation are required to protect this ecosystem. It may extend from the art of rewarding innovation to a legal system that protects property. There is a need for a social-economic environment that allows others to produce a favorable outcome. If we look at the stories of successful entrepreneurs across the world, it would become very apparent that many of them relied on breakthrough innovations to accumulate wealth. Some of the most prestigious universities can be given credit for nurturing future entrepreneurs by providing them with the knowledge of venture creation and helping them incubate their ideas. The process of entrepreneurship depends on the ecosystem supported by a political, social, and economic framework that can discourage or encourage venture creation.

The book recognizes the role of social acceptance and the role of women in fostering or reducing efforts for successful business creation. There is a need for a broad economic philosophy of jurisdiction that motivates these young venture capitalists to identify ways to maximize their economic value. The book argues that opportunities can be created even during a crisis like a pandemic and there is a need for creating a resilient ecosystem for these budding entrepreneurs.

Entrepreneurship is needed for the growth of individuals as well as nations. It is a logical strategy of the government to follow a path of identifying, nurturing, and building upon an environment where the seeds of successful venture creation can grow and flourish. Hopefully, the book can become a cornerstone in guiding future entrepreneurs in their journey toward the creation of successful business entities. The book offers a holistic perspective by contributing to current literature and practice of the importance of entrepreneurship.

Editors

Dr. Rajender Kumar is presently working in Department of Mechanical Engineering, FET, MRIIRS. He is actively involved and associated with various national level societies promoting Engineering/Technical/Quality concepts in India such as IEI (Member Local Chapter Committee), IIIE (Senior Member), QCFI (Governing Council Member), QCI (Registered Lead-Auditor), IIF (Member and Former Honorary Secretary, Faridabad Chapter), ISHRAE etc. He has received the grant from Government of India to conduct activities related to Entrepreneurship Development. He has 50 research papers and two book chapters to his credit. His area of expertise is Industrial Engineering and Management, Entrepreneurship and Business Incubation, SCM, Lean, Green and Agile Manufacturing System.

Dr. Rahul Sindhwani is presently doing post-doctoral research at IIM Amritsar. Prior to this, he was associated with the Mechanical Engineering Department, Amity University, Uttar Pradesh, India. He is an active lifetime member of the Indian Society of Technical Education and a fellow member of Institution of Engineering and Technology. He has teaching, research, and administrative experience of more than ten years. He has contributed more than 50 research papers/book chapters in the area of industrial engineering and management in various national/international journals of repute. His area of expertise is  industrial engineering and management, supply chain management, lean, green and agile manufacturing system along with analysis of problems using various optimization techniques.

Dr. Tavishi Tewary has over 13 years of experience in policy research and trade impact assessment. She has published various research papers in international journals of high repute. She has also conducted FDPs on Data Analysis. She has experience in providing scientific leadership to high-profile strategic sustainability and conservation initiatives with several years of research experience. She managed cross-sectoral and interdisciplinary teams of professionals to deliver on complex research projects. Currently, she is working as an assistant professor at Jaipuria Management Institute in Delhi NCR, Uttar Pradesh.

J. Paulo Davim earned his Ph.D. degree in mechanical engineering in 1997, M.Sc. degree in mechanical engineering (materials and manufacturing processes) in 1991, mechanical engineering degree (5 years) in 1986, from the University of Porto (FEUP), the Aggregate title (Full Habilitation) from the University of Coimbra in 2005 and the D.Sc. from London Metropolitan University in 2013. He is Senior Chartered Engineer by the Portuguese Institution of Engineers with an MBA and Specialist title in engineering and industrial management. He is also Eur Ing by FEANI-Brussels and Fellow (FIET) by IET-London. Currently, he is a professor at the Department of Mechanical Engineering of the University of Aveiro, Portugal. He has more than 30 years of teaching and research experience in manufacturing, materials, mechanical and industrial engineering, with special emphasis in machining and tribology. He also has interest in management, engineering education and higher education for sustainability. He has guided large numbers of postdoc, Ph.D. and master's students as well as has coordinated and participated in several financed research projects. He has received several scientific awards. He has worked as evaluator of projects for ERC-European Research Council and other international research agencies as well as examiner of Ph.D. thesis for many universities in different countries. He is the Editor in Chief of several international journals, Guest Editor of journals, books Editor, book Series Editor and Scientific Advisory for many international journals and conferences. Presently, he is an Editorial Board member of 30 international journals and acts as reviewer for more than 100 prestigious Web of Science journals. In addition, he has also published as editor (and co-editor) more than 125 books and as author (and co-author) more than 10 books, 80 book chapters, and 400 articles in journals and conferences (more than 250 articles in journals indexed in Web of Science core collection/h-index 54+/9000+ citations, SCOPUS/h-index 58+/11000+ citations, Google Scholar/h-index 75+/18500+).

Contributors

Harish Kumar Banga
Department of Fashion and Lifestyle
  Accessory Design
NIFT, Mumbai, India

Vikram Bansal
Atal Bihari Vajpayee School of
  Management and Entrepreneurship
JNU, New Delhi, India

Mahesh Chand
Department of Mechanical Engineering
J.C. Bose University of Science &
  Technology
Faridabad (Hr.), India

Dinesh Chawla
Department of Mechanical Engineering
FET, MRIIRS
Faridabad (Hr.), India

Kshitiz Choudhary
School of Business and Management
CHRIST (Deemed to be University)
Pune, India

Deepthi B.
Atal Bihari Vajpayee School of
  Management and Entrepreneurship
JNU, New Delhi, India

Parul Dubey
Department of Management
Pandit Sundarlal Sharma (Open) University
Chhattisgarh, India

Pushkar Dubey
Department of Management
Pandit Sundarlal Sharma (Open) University
Chhattisgarh, India

Prateek Gaur
Department of Management Studies
J.C. Bose University of Science &
  Technology
Faridabad (Hr.), India

Fr Jossy P. George
School of Business and Management
CHRIST (Deemed to be University)
Pune, India

Sanjeev Goyal
Department of Mechanical
  Engineering
J.C. Bose University of Science &
  Technology
Faridabad (Hr.), India

Shreya Homechaudhuri
Strategy, Innovation and
  Entrepreneurship
Institute of Management Technology
Ghaziabad (UP), India

Rajender Kumar
Department of Mechanical
  Engineering
FET, MRIIRS
Faridabad (Hr.), India

Vijay Kumar
Department of Applied Science
FET, MRIIRS
Faridabad (Hr.), India

Rupali Madan
Department of Management Studies
J.C. Bose University of Science &
  Technology
Faridabad (Hr.), India

Jayant Mahajan
School of Business and Management
CHRIST (Deemed to be University)
Pune, India

Om Prakash Mishra
Department of Mechanical Engineering
J.C. Bose University of Science &
    Technology
Faridabad (Hr.), India

Abhilash G. Nambudiri
Rajagiri Business School
Kochi, India

Vimlesh Kumar Ojha
Department of Mechanical Engineering
J.C. Bose University of Science &
    Technology
Faridabad (Hr.), India

Rajeev Saha
Department of Mechanical Engineering
J.C. Bose University of Science &
    Technology
Faridabad (Hr.), India

Kailesh Kumar Sahu
Department of Management
Pandit Sundarlal Sharma (Open) University
Chhattisgarh, India

Anshul Saxena
School of Business and Management
CHRIST (Deemed to be University)
Pune, India

Minakshi Sehrawat
School of Management and Liberal
    Studies
The NorthCap University
Gurugram, India

Deergha Sharma
School of Management and Liberal
    Studies
The NorthCap University
Gurugram, India

Rahul Sindhwani
Department of Mechanical
    Engineering
Amity University
Noida (UP), India

Sushant
Department of Mechanical
    Engineering
FET, MRIIRS
Faridabad (Hr.), India

Shalini Rahul Tiwari
Strategy, Innovation and
    Entrepreneurship
Institute of Management
    Technology
Ghaziabad (UP), India

Srijna Verma
Department of Management Studies
J.C. Bose University of Science &
    Technology
Faridabad (Hr.), India

1 Lean in Business Models

Way to Go for Entrepreneurial Journey in Industry 4.0 Context

Rajender Kumar, Dinesh Chawla, Sushant,
Rahul Sindhwani and Harish Kumar Banga

CONTENTS

1.1 Introduction..1
1.2 Literature Review...3
    1.2.1 Start-ups..4
1.3 Lean Manufacturing: Concept & Beyond...6
1.4 Lean Start-Ups: Concept & Beyond 8
1.5 Relationship among Lean Manufacturing and Lean Start-Ups 13
1.6 Barriers to Implementing Lean in Start-Ups 13
1.7 Conclusion 15
References...16

1.1 INTRODUCTION

The fundamental objective of any business is that the product must reach the customer within the stipulated time frame (favorably less) and the consumer must capitalize on the maximum benefits out from the products/services (within the product shelf life) (Merritt, 1974; Gemmell, 2017). In the earlier age of industrialization, the availability of resources was limited and resulted in creating a gap among the demand and supply of goods/services (Kumar et al., 2018). At that time, it was popular to learn from mistakes because prediction or forecasting was the toughest task. Eventually, this huge gap creates the dealt situations for the industries to perform under the loss condition (Drucker, 2014). In addition, the industries were not in the position to have better control on the operations as well as losing their fruitful resources. That was the main reason behind the fewer industries in operation and only those entrepreneurs having the capability to bear risks were beginners of an entrepreneurial journey (Müller et al., 2018). Even, being an entrepreneur remains a hot topic for discussion and research. It is an exaggeration to say that to become an entrepreneur, hundreds of motivations attract human beings to start their entrepreneurial journey. Figure 1.1 represents the model for motivations to the persons to be an entrepreneur in practice since 1986.

DOI: 10.1201/9781003256663-1

1

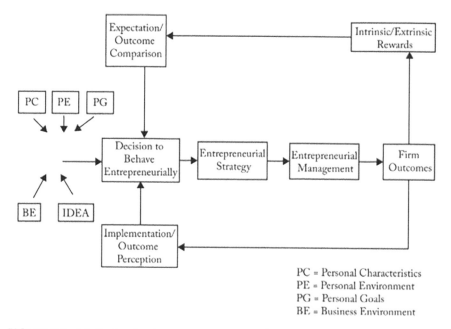

FIGURE 1.1   Motivation for the entrepreneurs to go for entrepreneurial journey (Naffziger et al., 1994).

Entrepreneurs begin their journey with devotion and motivation for initiating a start-up (Bienzeisler, 2008). The term entrepreneur means "to-undertake." In a simplified manner, the entrepreneur has the execution plan. The entrepreneurial process begins with identifying (creating/seizing) the opportunity and continuing working hard for the conversion of these opportunities into reality regardless of the resources currently controlled (Ries, 2011). For this purpose, an enterprise pays a certain amount for procuring the resources (available in the market) and thereafter converts those into useful products/services to cater to the end customer's needs. The initial investment done by the enterprise on the products and services is taken by selling the products at an uncertain price. Additionally, the entrepreneur bears all the risks with the business/enterprise (Ferreira and Lisboa, 2019). The present-day entrepreneurs have a focus on delivering the services/products to the customer/society by organizing all the resources in such a systematic manner that end customers should feel privileged (Wahl and Munch, 2021).

In the present context, most industries are looking for Industry 4.0 to have a better competitive advantage. That's why the entrepreneurs are now comparatively viewed as the practitioners instead of risk bearers in the past (Gans et al., 2019). The most important task for the present-day entrepreneur is to create value for the customers that would help his start-ups/entrepreneurial setups not only for the survival in a longer period of services but also for thriving in a short period (Borland et al., 2018). For this, each entrepreneur develops their unique models based on pre-assumed conditions for creating such values and crafts experiments to test/evaluate their hypotheses. The published literature on industrialization also reveals that securing a highly motivated work environment is not an accident (Henderson and Venkatraman, 1993). It is always the result of systematic and concerted efforts of management to improve and create a system of motivators. For motivating the workforce, holistic approaches are used (Safar et al., 2018). Implementing lean is also one of them.

From the past decade, the most influential as well as hot topics for discussion among academicians and researchers is Industry 4.0 (I-4). I-4 aims to manage the business activities in effective and efficient manner whereas the start-ups relate to initiation of the business. I-4 in general supports an industry to make the right decisions by providing the complete information regarding particular situation through the effective utility of technology. Not only for effective decisions, I-4 also contributes in automation of resources to fulfill consumer demand. The advance technological tools/techniques like artificial intelligence, blockchain technology, data-analytics, machine learning etc. are used for developing and building viable models for running the industries. These techniques help in performing all the business operations beginning from demand identification to the delivery of the desired product to the user end. The various benefits from Industry 4.0 are as follows:

a. Performance enhancement
b. Efficacy and effectiveness
c. Meeting compliance
d. Customer-driven
e. Operational-cost
f. Effective-communication
g. Opportunities for future developments
h. Effective decision-making
i. Agility in operations
j. Increased profit and ROI

The present study will emphasize understanding the unintended consequences at the beginning of an entrepreneurial journey and how they can be tackled easily without impact on an entrepreneur's motivation and business style. In addition, the study will highlight the importance of lean in business models. The expected benefits of implementing lean in the business model are as follows:

a. Conceiving the consumer requisites
b. Creating value for the customer
c. Collaborating with stakeholders for specialized services
d. Capital management (wealth management)
e. Competing in national/international clusters

1.2 LITERATURE REVIEW

The motivation for an entrepreneur lays the foundations of encouragement which provide the basic platform to perform and achieve the desired goals (Ardianti and Inggrid, 2018). Each entrepreneur is motivated by a variety of reasons such as money, brand, reputation, skill, experience, and many more (Shahab et al., 2019). Most young people join the entrepreneurial journey just for the sake of creating wealth. They are highly motivated and keep the tagline in their mindsets i.e., no need to differentiate between what you have sought and what happens actually after bearing the risk, it's time to learn from the experiences (Roy, 2011). That's why, most of the start-ups are failing within the first or second year of the journey and the same has been depicted from national economic surveys. Even in some situations, it has been observed that few well-designed/tailor-made products are not accepted by the market. Start-ups are generally crafted for valuing the end customer.

At the early stage, start-ups were mis-conceptualized and considered small business entities. The scalability factor helps in differentiating among start-ups and small business entities (Maurya, 2018). The entrepreneurial setups operating based on a flexible business model to scale the growth through effectively utilizing the technology are referred to as the start-ups. On the other hand, small business entities are working on fixed business models (Smith et al., 2010; Ries, 2017).

Both lean manufacturing and lean start-up have gained attention in the past two decades. The inclusion of the lean concept focused on doing more with lesser efforts. It helps in getting the things/functions done right on the very first attempt (Euchner, 2021). In the following subsections, the concept of lean manufacturing and lean start-up is explored for better insights.

1.2.1 START-UPS

In the start-up perspective, the consumer's perspective on expectation (what he is expecting); product (what he is getting), and the experience (user-experience with the product/services) are the most important concerns (Cohan, 2018). These concerns if handled carefully will lead to strengthening the start-up's core competencies and ultimately success (Ardito et al., 2015). The vision to go on an entrepreneurial journey always plays an important role. It helps in defining the cause to which the start-up is going to fulfill and also, highlights the outcomes that reveal the end product with a variety of characteristics (Lechner and Gudmundsson, 2012).

For initiating start-ups, the entire entrepreneur is generally working on different business models like business to business; business to customers simultaneously, etc. At the initial stage it is too difficult for them to go with a single business model that can meet all the desires of both consumers and entrepreneurs (Zarezankova-Potevska, 2021). For finalizing any business model, the validation of the model is done where a lot of resources are wasted for getting the desired product/services (Spender et al., 2017). Also, the implementation of such crafted values in the products/services/process requires a systematic approach because it has direct impacts on the start-up performance (Bakator et al., 2018). The possible start-up's success enablers for the start-ups are as follows:

1. **Viability of idea:** In an idea, the two most important facts are conceiving and convincing. Here, conceiving refers to the person's capability to understand the aspects behind the idea to pursue i.e., purpose-driven based on situation, product/process, place, etc. (Subrahmanya, 2021). The convincing refers to the

visibility and viability of the idea which is again factor-driven. The idea that reveals both conceiving and convincing will surely lead to success.

2. **Funding Sources:** A very common hurdle for all entrepreneurs i.e., availability of funds. The earlier studies on findings to start-ups reveal that at the early stage the entrepreneurs should go for their own money or if required, use the funds from relatives or the bootstrapped funds (Kumar et al., 2021). These amounts should be used for the conversion of ideas into reality. Once the idea gets validation, the entrepreneur can go for seed fundings from angel investors or private firms (BR et al., 2018; Dutta et al., 2021). The studies reveal that more than 50% of start-ups failed to get seed funding just because of not validating the ideas. For upscaling the business, inviting venture capitalists, loans from the bank, etc. is the better choice (Singh et al., 2022; Dutta et al., 2022). All kinds of funding sources are having their terms and conditions and possibly may cause the start-ups to function. So, while looking for funds, it becomes prominent for an entrepreneur to understand the need for funds, funding sources available, and the reason behind funding carefully (Bilan et al., 2019).

3. **Resources accumulation and utilization:** The success story of all big entrepreneurs reveals that they have a very limited number of resources at the initial stage. This may be the possible cause for failure, but they effectively manage the resources and get success. For entrepreneurs, the accumulation of the right kind of resources at the right time with pricing to meet their objectives is very important (Rosli and Chang, 2020). Even, the utility of those resources should be done in an optimized manner to get better outcomes.

4. **Target Market**: While looking at the business entities, the main factor behind failures is the target market where the entity desires to sell the products/services. For the start-ups the identification of the target market is important and this should be done carefully (Juhdi et al., 2015; Sindhwani, 2022b). The start-ups should work on distinguishable aspects to grab the market i.e., differentiating among products/services characteristics over competitors; cause behind serving; clear mission and vision demonstration; target customer categories etc. (Yavuztürk et al., 2019).

5. **Technological aspect**: Technology development plays a crucial role in start-up functioning. The deployment of technology leads to skilling and striving for excellence for the start-ups (Johnson, 2006; Sindhwani et al., 2022a). Technological advancement can contribute directly to achieving the desired outcomes such as performance enhancement (efficacy and effectiveness); making it customer-driven; increasing profit and return on investments; reducing the overall operational costs; communicating effectively among all stakeholders and making the decision process more effective, etc.

Despite the success stories, there are issues with the decision process in start-ups that delimit the growth. The common issues related to the decision process are listed below:

- Regarding the inventory (accumulation, holding, utility)
- Regarding the framework and synchronization among value-chain

- Regarding communication among stakeholders (supply, demand, product features, pricing, promotion)
- Regarding the utility of resources based on push or pull mechanisms
- Regarding the nature of business and target market (Robertson, 2003; Petru et al., 2019)

1.3   LEAN MANUFACTURING: CONCEPT & BEYOND

During the twentieth century, the most typical tasks for all industries were resources identification and accumulation. Besides that, it was also difficult to create a single plate-form where all stakeholders come together and contribute (Gadenne, 1998; Kumar et al., 2016). That's why, the earlier studies on industries reported a gap among consumer demand and product/service deliveries (Kumar et al., 2017; Chang and Cheng, 2019). Eventually, this huge gap creates situations for the industries to perform under the loss condition (Shah and Ward, 2007; Kumar et al., 2014a). At that time, it was popular to learn from the mistakes because prediction or forecasting was the toughest task. Industries were not in the position to have better control over the operations as well as losing their fruitful resources.

During that period, lean management came into existence. The concept of lean management (also known as lean methodology/lean terminology/lean principles) was firstly coined for implementation in the automobile manufacturing sector. The objective behind implementing lean was to manage the execution part smoothly and that was the moment that triggered the shifts in the manufacturing paradigm to avoid/eliminate waste (Shanker et al., 2019; Womack et al., 1990). The unintended waste associated with the production process directly impact the efficiency and effectiveness which further caused customer satisfaction (Feld, 2000; Sindhwani et al., 2021a).

Lean in general, taking care of the timelines of the product, information, and cash flow from one to another and vice-versa (Kumar et al., 2014). The initiation of lean begins with consumer need identification and ultimately to better user experiences (Nagar et al., 2021; Shah and Ward, 2003; Sindhwani et al., 2021b). It generally works on identifying and eliminating waste such as inventory (over and under); processes (more or less than required); production quantity (over and under), unnecessary transportation, etc. (Reichhart and Holweg, 2007). While discussing lean, it is important to highlight the principles which provide the clear roadmap regarding the lean implementation (Singh et al., 2019). There are five distinct principles: a) consumer's need identification; b) understanding the existing value-stream to differentiate among value vs non-value added processes; c) defining and demonstrating the changes in product flow; d) creating the customer pull; and e) pursuing perfection to sustain the competitive advantage represented in Figure 1.2 (Womack, 1990).

The studies on lean implementation in the manufacturing context reveal the benefits like clarity in setting the objectives/goals; optimized resource consumption; standardized resource flow; customized process flow; effective and efficient team-building; and ultimately production cost reduction, etc. (Mittal et al., 2016). Lean manufacturing has the objectives to eliminate the waste not only to the production line but also include the customer and supplier network management. Lean manufacturing emphasizes lesser incorporation in terms of human efforts, inventory, time consumption in production

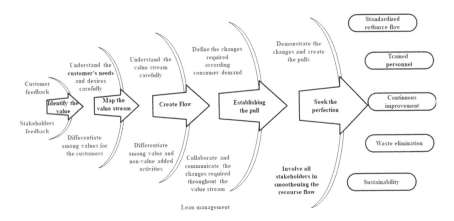

FIGURE 1.2   Lean principles.

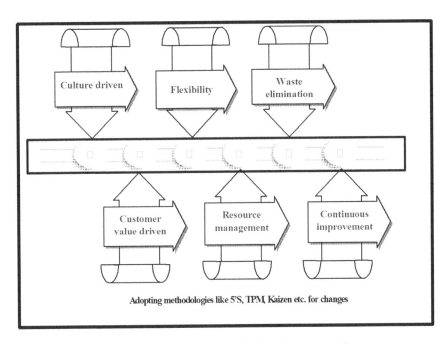

FIGURE 1.3   Factors contributing to the success for lean implementation.

and delivery processes and even, the space available. It would result in enabling the organization to be highly responsive towards consumer demands. The success factors of implementing lean management are (represented in Figure 1.3).

Since its inception, almost four decades have passed; still, this concept is more useful in the manufacturing sector in comparison to the service sector (Mittal et al., 2017). The main hurdle in the application of lean in the service sector is real-time decision-making in an uncertain environment. Even, the identification and stratification for sub-grouping purposes to have the real time data within a short period is a difficult task.

1.4 LEAN START-UPS: CONCEPT & BEYOND

In earlier days, the groundwork for initiating any business entity was necessary such as the product demand, material, specification, and supplier identification. Nowadays, the widely diffused approach named lean methodology in business model development of start-ups is proposed by the practitioners (Ries, 2017). The application of the lean start-up model in the Industry 4.0 context seeks to implement the scientific approach that helps an entrepreneur to undertake the customer value priority by structuring the assumptions lying very close to the consumer expectations (Euchner and Blank, 2021). In lean integrated models, consumer feedback (before and after utilization of product/service offered by the industries) is directly involved in framing the policies/strategies to streamline the start-up operations (Buchalcevová and Mysliveček, 2016). It would help in creating zeal and enthusiasm among the workforce by eliminating the waste and further streamlining with the enterprise objective to have both tangible and intangible benefits. The key values that are provided to the customers by the lean start-ups are represented in Figure 1.4.

The involvement of lean has provided several tools/techniques which are useful in start-up operations such as business modeling, customizing the goods/services based on minimum viability, consumer identification, and many more (Bhatti and Zia, 2021). The utility of lean in the start-up model is the important key curricular foundation and is attracted by most entrepreneurs (Wirtz, 2016). The reason behind the attraction and involvement of lean in the start-up models is the elimination of waste activities from the business models and creating values for the end consumers. The lean canvas model is represented in Figure 1.5.

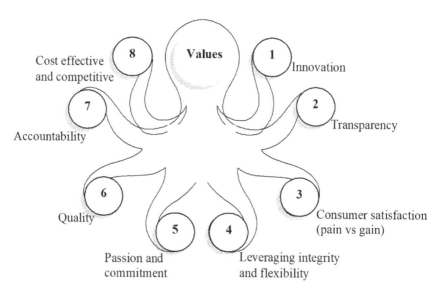

FIGURE 1.4 Key values that customers are looking for.

1. Problem: List the customer pains and based on prioritization choose three important problems	4. Solution: List the customer gains clearly	3. Unique value proposition: Crafting the values carefully	9. Unfair advantage: Report the unfair advantage that can't be copied easily	2. Customer Segments: List the targeted customer segment and market niche if any
Existing alternatives: If there is any alternative available, report the studies on strengths and weaknesses of the existing alternatives	8. Key metrics: List the key members in terms of acquisition, activation, retention, revenue, referral	High-level concept: Report the uniqueness as it matters for having the competitive advantage over the others	5. Channels: Strongly mention the channels to connect with end-customers directly	Early adapters: List the early adapters based on factors related to problem to be solved

6. Cost structure: List the various components of operational costs such as fixed cost, variable cost, office rent, administrative cost etc.	7. Revenue stream: List the ways for getting the revenues

Break-Even

FIGURE 1.5 Lean Business Canvass (Maurya, 2012).

STEP 1: LIST THE PROBLEMS

It is essential for the start-ups that they should identify customer pain areas carefully. Lean start-ups mainly emphasize listing the three top-most problems for the initiation of any start-up (Teece, 2010). The case studies of most of the various start-ups reveal that there is a significant relationship between the list of problems and having interdependency. Still, there are possible barriers in identifying the problems:

1. Understanding the pain of the consumers as individuals and as a group is not an easy task. It requires the systematic evaluation of market trends, alternatives analysis, the feasibility of features/characteristics to be added, and many more.
2. Listing out or setting the priority to the problems when targeting the customers is another barrier. The charting process for the problems should be logical as well as feasible so that the targeted audience can meet the required products/services.
3. The presence of alternatives in the market also creates issues for the start-ups in listing out the problems (Felin et al., 2020).

Here, the use of advanced technology-based tools can help the start-up to connect with the consumers directly. Even, in a few start-ups, it provides the specific support to generalize the hypothetical situation. In addition, the dedication, conviction, and devotion of the entrepreneur will also help in pointing out the problems (Yang et al., 2018).

STEP 2: TARGETING AUDIENCE

From the initial days of the industrial revolution, the customer is the only boss for any business. In the lean start-up model, both the problem identification and defining the market segment (which is to be targeted) are done simultaneously (Ghezzi and Cavallo, 2020). As stated in the problem identification section, the problem identification and finalization need a systematic evaluation to be done. Here, it is an exaggeration to say that this systematic evaluation should be based on customer-centricity (Blank and Dorf, 2020). Aligning the customers to problems is to be done in a manner that the customers will contribute to start-ups in finding the solution. Especially, the involvement of early adopters actively in listing specific problems will surely help in start-up initially then in scaling.

STEP 3: UNIQUE VALUE PROPOSITION

Nowadays, the most important factor which can help in deciding whether the start-ups will succeed or fail in the future is the unique value proposition (Santisteban et al., 2021). It is a well-known fact about the product/services that the customer forgets the prices but they remember the quality of the product for a long time (Alvarez, 2017). In today's scenario, it becomes prominent to understand the unique selling proposition (USP) which helps in defining the statements for differentiating the start-up from its competitors. The value proposition to the customer should be aligned with the business drivers such as product/service characteristics, cost factor, compliances, customer acquisition and retention strategies, turnaround time in fulfilling the demand, profit, etc. (Walsh, 2009). While specifying the value to the customer, the entrepreneurs always initiate the value in terms of maximizing (profit/productivity/services) vs minimizing (wastes/processes/timings); increasing (efficiency/effectiveness) vs decreasing (production and delivery costs), etc. While proposing the value, the six interlinked and distinct elements are as follows:

1. *Market:* Represents the targeted group
2. *Value:* Represents the benefits that the customers are looking for
3. *Offerings:* Represents the kind of product and service offered to the consumer as an individual or in a mix.
4. *Benefits:* Represents the delivery of values to the consumers
5. *Differentiation:* Represents the competitive advantage over the alternatives
6. *Proof:* Represents the credibility through the selling trends and customer feedback

STEP 4: DEFINING THE SOLUTIONS

Based on the problem listed in Step 1, the start-ups need to list the solutions for the problems. Here again, point out the topmost top issues (problems) which are being resolved is to be represented. In general, the product/services characteristics that meet the minimum viability criteria are listed to attract the crowd (Harms and Schwery, 2019). In start-ups, the products with minimal features and maximum utility are innovated. The list of principles for designing and developing the minimum viable products is represented in Table 1.1.

STEP 5: CHANNELS TO CONNECT WITH CUSTOMERS

Most of the start-ups are failing day to day because they are not building a significant path to connect with the customers. In earlier days, it was difficult to connect with the customers due to the requisition of more and more resources. But now, with the technological revolution, the start-ups can easily connect with consumers and channel the flow of resources, especially the information flow (Jain and Khandelwal, 2020). The common channels used widely are free channels

TABLE 1.1
MVP Principles

Sr. No.	Principle-	Description
1	Less is more in MVP development	The first and foremost principle for MVP is the lesser features/characteristics with maximum utility.
2	The most boring feature is the most important one	The second principle is related to have the advantage over competitors by recognizing the specific features and providing the solution for them.
3	The path of least resistance: minimize friction	In the MVPs, the waste should be eliminated at the design stage only.
4	Think scarcity: demand vs supply	The products/services should bridge the gaps among demand and supply.
5	Focus on closing the loop for the users	Start-ups should design and develop the products/services to cater the need of the consumer and be flexible enough to change according to feedback.
6	Iterate and refine with user feedback and usage data	The before and after buying feedback of the customers will enable the start-ups to go for the requisite changes.
7	Be prepared for pivoting if it's not working	Always, the alternatives strategies/plans/ products/processes must be in the queue because if the primary action fails, the secondary will play the lead role in penetration process.

(social media platforms); paid channels (search engine marketing); inbound marketing (pull messaging); outbound marketing (push messaging); communication done manually; automated communication; directly to individuals and indirectly through the partners etc. (Zollo et al., 2016).

STEP 6: COST STRUCTURE

For start-ups, the cost incurred on resources plays an important role. It will help in deciding the capacity and the capability of the start-ups to stand the market requisitions. There are direct and indirect costs involved in the start-ups i.e., fixed cost; variable cost and administration cost etc. All these costs are interrelated and contribute to the start-ups from the beginning throughout the journey. The complete and concussed breakups for all these costs should be done carefully and depicted in the model.

STEP 7: REVENUE STREAM

The revenue stream represents the various ways to get revenues for a start-up. It generally comprises revenues based on transactions and recurring in the projects/services of the start-ups. The revenue can be in the terms of profits, acquisition, equity, licensing share, after-sale servicing, consultation and brokerage, advertising, etc. A start-up needs to align the cost structure with the revenue stream for the effective and efficient utilization of financial resources to build a sustainable start-up.

STEP 8: DEFINING AND DEVELOPING THE KEY METRICS

At the initial state of the start-ups, it is essential to find and list the key contributors that can help in providing the required information about the business in real-time before the sales report comes. In the lean start-up model, the key metrics are designed considering the following terms

1. **Acquisition:** Represents the number of ways to acquire the consumers using the channels mentioned in step-5.
2. **Activation:** Represents the willingness of potential customers to go for the services initially.
3. **Retention:** Represents the continual of services by the customers in future aspects
4. **Revenue:** Represents the revenue earned by selling products/services to the customers
5. **Referral:** Represents the increase in the count of potential buyers based on references by the existing satisfied customers

STEP 9: DEFINING THE UNFAIR ADVANTAGE

In the competitive environment, all the start-ups need to have an unfair advantage. Generally, the unfair advantage focused on having the legal rights against the

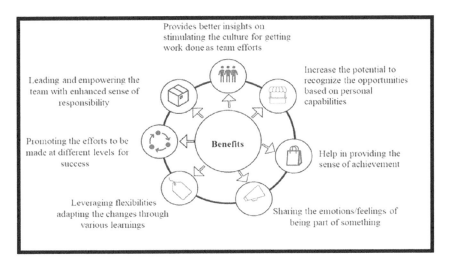

FIGURE 1.6   Steps to promote the lean culture.

properties (tangible or intangible) intellectually. This further limits the competitors in copying or duplicating the products/processes subsequently. Nowadays, the unfair advantage also considers the aspects like number of potential investors, brand royalty, networks, community, etc. Figure 1.6 reveals the benefits of lean start-ups.

1.5   RELATIONSHIP AMONG LEAN MANUFACTURING AND LEAN START-UPS

Lean in business model and manufacturing sector initiating with the concept of cultural development where all the stakeholders are motivated to contribute according to the competence in streamlining the operations (Ghezzi et al., 2015; Isensee et al., 2020). Further, the stimulation of culture helps in identifying and recognizing the individual operation/activity requisites and crafting the policies that suit all stakeholders best. In the Table 1.2, the relationship between principles of lean manufacturing and start-ups is elaborated to reveal the commonality among both two concepts.

The deployment of lean will provide a sense of understanding regarding the desired objectives and also the achieving strategies (Link, 2016). Most of the issues are resolved strategically such as spotting the activity of the waste, formulating the solution, inviting collaborations, etc. and the continuous improvement is done based on the learnings of past events.

1.6   BARRIERS TO IMPLEMENTING LEAN IN START-UPS

Here, it is an exaggeration to say that listing out the components of all nine activities is very crucial. Most of the entrepreneurs at present do not have a clear understanding of all activities of the business model. Even, they have failed to establish a healthier

TABLE 1.2

Relationship among Lean Manufacturing and Lean Start-ups

Lean manufacturing principles	Lean start-up principles	Similarity
Specify Value: Lean manufacturing initiates with the value identification i.e., what the customer wants exactly?	*Entrepreneurs are everywhere*: Lean start-up initiates with the identification and recognition to an entrepreneur based on his motivation, innovation and conviction in to creating the products/services under extreme uncertain environment.	No-matter how big the industry size is, the concept of lean can be applied anywhere.
Identify the Value Stream: In the second principle, differentiation among value-added and non-value-added activities is done which is a very crucial task.	*Entrepreneurship is management:* Managing the innovation plays a significant role in initiating and growth of the start-ups through differentiation from alternatives.	A certain skill-set and knowledge is required to undergo the customer values systematically and fulfill the value strategically.
Make the Product Flow: Eliminate the non-value-added activities from the value chain and manage the uninterrupted flow of resources.	*Validated learning:* For the start-ups it is essential to have the sustainable value chain which specifies the vision of the start-ups and gets validated based on logical facts and findings.	Experimentation is done to validate to justify the gain areas vs the pain areas.
Customer Pull: In the industries, creating the customer pull provides the competitive advantage through interlinking all the activities beginning from demand to supply.	*Build-Measure-Learn:* Start-ups must focus on building the solution based on consumer perceptions.	Helps in pivoting or persevering to the consumer demand especially when the end user is involved in dictating the products/services attributes
Pursue the Perfection: Pursuing perfection means continuously improving and sustaining the achieved results in the long-term perspective.	*Innovation accounting:* Continuously improve the start-ups processes and services effectiveness by holding the innovators accountable.	The process owners must be accountable for sustaining the process outcomes.

relationship among these activities which cause the start-up's real-time functioning. The barriers in implementing lean in start-ups are as follows:

A. **Implementing lean manufacturing in start-ups:** Lean is simply about creating more value for customers through eliminating the activities that do not contribute any value to products/services. The systematic approaches are used to identify and eliminate non-value-added activities from the key operational areas (Mc-Elwee and Frith, 2008). The existing literature available on lean manufacturing reveals that this methodology has a dependency on culture and is explicitly implemented for the continuous/incremental improvements in the existing processes/products. For the initial tangible outcomes after lean implementation, an industry has to wait for a minimum of three years. The initial cost of implementing lean is also higher [Womack et al., 1990]. The lean manufacturing approach is now used for start-up business models. Likewise, the manufacturing industry performance enhancement, it is firmly believed that lean would also contribute in accelerating the start-ups striving and sustaining effectively. It would surely contribute to the innovation process as it is the customer-centric approach and also, help in eliminating the non-value-added activities in the business model itself [Ries, 2011]. As far as a reality for start-ups is concerned, it has been revealed that there is a limitation of resources, especially initial fundings. Even, no entrepreneur can devote all resources and wait for more time to get the initial results. Also, for the start-ups, it is important to create significant value through radical innovations then the incremental improvements in existing products.

B. **Validation of customers:** Nowadays, the traditional business plans are dominated and discouraged by the lean start-up models. In lean start-ups, the healthier channel for communication among the consumer and producers is necessary to build for iterating the consumer feedback and learning from them (Masudin et al., 2021). In the lean start-up model, importance is given to the design and development of minimum viable products instead of wasting time interacting with customers. Here, the customers interact to get feedback about the product/services as input for improvement [Dorf and Blank, 2012]. In the present-day start-ups, the focus is given on emotional marketing for attracting the premier customer lobby, especially the customized products/services.

For any business entity, customer validation is considered as the best way to eliminate the additional efforts for finding the markets for finished products. Though, it is not the best practice for all types of business because of dependency on various factors such as product characteristics; consumer perception, etc. Even the customer validation itself is a huge task and requires a huge investment. Most of the young entrepreneurial setups fail just because of working on the customer validation part.

C. **Lack of awareness on lean canvas:** In lean start-ups, the development of a lean business model canvas is the popular tool and shown in Figure 1.4. The figure reports the nine distinct activities and their descriptions that need to be defined by the entrepreneurs [Osterwalder and Pigneur, 2010]. The objective behind representing these activities is to provide a clear understanding to all the stakeholders on a business model. All the activities are interrelated with each other and can impact the start-up's functioning if not planned systematically. Also, many experimentation works are done among the activities before finalizing the model.

1.7   CONCLUSION

The lean concept not only can be implemented in the manufacturing sector but also the start-up business models. It simply specifies the importance of culture that further responds to the non-value-added activities effectively and efficiently. It helps in accelerating the performance of respective entities whether it belongs to manufacturing or the start-up's community. Lean focuses on getting optimized outcomes through the optimum utility of the resources.

The present study explores the opportunities for present-day entrepreneurs by understanding the lean concept in brief. Even, the justification among lean manufacturing and lean start-ups is discussed for providing an insight to focused areas for lean implementation. The concept of lean in general helps an entrepreneur in executing the desired activities through scientific and experimentation-oriented approaches. It empowers the entrepreneurs by providing support in decision-making based on the data available. The implementation of lean in start-ups can contribute to building blocks towards consumer desires and satisfaction. The strategic roles which are mostly looked at by the entrepreneurs from the lean implementation are:

1. Dealing in an efficient and effective manner with all kinds of operations
2. Adequate support in decision-making
3. Interlinking the start-up strategic policies with the objectives
4. Exploring and analyzing the business environment concerning opportunities and risks
5. Listing the key contributors
6. Suggesting ways to deal with risks or alternatives

Despite the useful gains from the lean concept implementation, there are limitations also. The literature available on lean manufacturing offers less guidance for the aspirants to go on an entrepreneurial journey using lean start-up model especially in creating the new products radically and the markets. In addition, the biggest barriers for start-ups at the initial stage i.e., development of minimum viable products and the customer validations are not addressed very well. There is no clue regarding both the barriers i.e., when and why to go for both the two? As far as the literature on lean start-ups is concerned, a small mistake can lead to failure at any stage in future developments. So, it is recommended that the start-ups should prepare the strategies based on alternative streams for experimentation in their commitments to the consumers.

REFERENCES

Alvarez, C. (2017). Lean customer development: Building products your customers will buy. O'Reilly Media.

Ardianti, R. and Inggrid, N. (2018). Entrepreneurial motivation and entrepreneurial leadership of entrepreneurs: Evidence from the formal and informal economies. *International Journal of Entrepreneurship and Small Business*, 33(2), 159.

Ardito, L., Messeni Petruzelli, A. and Albino, V. (2015). From technological inventions to new products: a systematic review and research agenda of the main enabling factors. *European Management Review*, 12(3), 113–147.

Bakator, M., Đorđević, D., Ćoćkalo, D., Nikolić, M. and Vorkapić, M. (2018). Lean startups with industry 4.0 technologies: Overcoming the challenges of youth entrepreneurship in Serbia. *Journal of Engineering Management and Competitiveness (JEMC)*, 8(2), 89–101.

Bhatti, S. H. and Zia, M. M. (2021). Lean startup approaches and business model innovation—A case study on digital startups in an institutional void. In *Business model innovation: New frontiers and perspectives*. Routledge, 40–62.

Bienzeisler, B. (2008). Business transformation: New organizational and business models. In D. Spath & W. Ganz (Eds.), *The future of services: Trends and perspectives* (pp. 247–264). München: Hanser. Fraunhofer-Publica.

Bilan, Y., Đšuzmenko, Đ. and Boiko, A. (2019). Research on the impact of industry 4.0 on entrepreneurship in various countries worldwide. In Proceedings of the 33rd International Business Information Management Association Conference, IBIMA 2019: Education Excellence and Innovation Management through Vision 2020, 2373–2384.

Blank, S. and Dorf, B. (2020). *The startup owner's manual: The step-by-step guide for building a great company*. John Wiley & Sons.

BR, G. K., Majumdar, S. K. and Menon, S. (2018). Manoeuvre of electronic entrepreneurial ecosystem to contemporary indicator of techno business leadership in industry 4.0: Digital entrepreneurship. *Global Journal of Enterprise Information System*, 10(3), 25–33.

Borland, H., Lindgreen, A., Maon, F., Ambrosini, V., Florencio, B. P. and Vanhamme, J. (2018). *Business strategies for sustainability*. Routledge.

Buchalcevová, A. and Mysliveček, T. (2016). Lean startup and lean canvas using for innovative product development. *Acta Informatica Pragensia*, 5(1), 18–33.

Chang, A. and Cheng, Y. (2019). Analysis model of the sustainability development of manufacturing small and medium- sized enterprises in Taiwan. *Journal of Cleaner Production*, 207, 458–473.

Cohan, P. S. (2018). Boosting your startup common. In *Startup cities*. Apress, 219–235.

Dorf, B. and Blank, S. (2012). *Startup owner's manual: The step-by-step guide for building a great company*. K & S Ranch.

Drucker, P. (2014). *Innovation and entrepreneurship*. Routledge.

Dutta, G., Kumar, R., Sindhwani, R. and Singh, R. K. (2021). Digitalization priorities of quality control processes for SMEs: A conceptual study in perspective of Industry 4.0 adoption. *Journal of Intelligent Manufacturing*, 32(6), 1679–1698.

Dutta, G., Kumar, R., Sindhwani, R. and Singh, R. K. (2022). Overcoming the barriers of effective implementation of manufacturing execution system in pursuit of smart manufacturing in SMEs. *Procedia Computer Science*, 200, 820–832.

Euchner, J. A. (2021). *Introducing a new business model: The business model pyramid*. Lean Startup in Large Organizations, 75–98.

Euchner, J. and Blank, S. (2021). Lean startup and corporate innovation. *Research-Technology Management*, 64(5), 11–17.

Feld, W. M. (2000). *Lean manufacturing: Tools, techniques, and how to use them*. CRC Press.

Felin, T., Gambardella, A., Stern, S. and Zenger, T. (2020). Lean startup and the business model: Experimentation revisited. *Long Range Planning*, 53(4), 101889.

Ferreira, V. and Lisboa, A. (2019). Innovation and entrepreneurship: From schumpeter to industry 4.0. *Applied Mechanics and Materials*, 890, 174–180.

Gadenne, D. (1998). Critical success factors for small business: An inter-industry comparison. *International Small Business Journal: Researching Entrepreneurship*, 17(1), 36–56.

Gans, J. S., Stern, S. and Wu, J. (2019). Foundations of entrepreneurial strategy. *Strategic Management Journal*, 40(5), 736–756.

Gemmell, R. M. (2017). Learning styles of entrepreneurs in knowledge-intensive industries. *International Journal of Entrepreneurial Behavior & Research*, 23(3), 446–464.

Ghezzi, A. and Cavallo, A. (2020). Agile business model innovation in digital entrepreneurship: Lean startup approaches. *Journal of Business Research*, 110, 519–537.

Ghezzi, A., Cavallaro, A., Rangone, A. and Balocco, R. (2015). A comparative study on the impact of business model design & lean startup approach versus traditional business plan on mobile startups performance. Proceedings of the 17th International Conference on Enterprise Information Systems.

Harms, R. and Schwery, M. (2019). Lean startup: Operationalizing lean startup capability and testing its performance implications. *Journal of Small Business Management*, 58(1), 200–223.

Henderson, J. C. and Venkatraman, H. (1993). Strategic alignment: Leveraging information technology for transforming organizations. *IBM Systems Journal*, 32(1), 472–484.

Isensee, C., Teuteberg, F., Griese, K. and Topi, C. (2020). The relationship between organizational culture, sustainability, and digitalization in SMEs: A systematic review. *Journal of Cleaner Production*, 275, 122944.

Jain, R. and Khandelwal, R. (2020). Dare to defy the challenges of online business. *International Journal of Entrepreneurship and Innovation Management*, 24(4/5), 281.

Johnson, P. (2006). Business models. In P. Johnson (Ed.), *Astute competition (Technology, innovation, entrepreneurship and competitive strategy, Vol. 11)*. Emerald Group Publishing Limited, 53–72.

Juhdi, N. H., Hong, T. S. and Juhdi, N. (2015). Market orientation and entrepreneurial success: Mediating role of entrepreneurial learning intensity. *Jurnal Pengurusan*, 43, 27–36.

Kumar, R., Kumar, V. and Singh, S. (2014). Effect of lean principles on organizational efficiency. *Applied Mechanics and Materials*, 592–594, 2613–2618.

Kumar, R., Kumar, V. and Singh, S. (2014a). Role of lean manufacturing and supply chain characteristics in accessing the manufacturing performance. *Uncertain Supply Chain Management*, 2(4), 219–228.

Kumar, R., Kumar, V. and Singh, S. (2016). Relationship establishment between lean manufacturing and supply chain characteristics to study the impact on organisational performance using SEM approach. *International Journal of Value Chain Management*, 7(4), 352.

Kumar, R., Kumar, V. and Singh, S. (2017). Modeling and analyzing the impact of lean principles on organizational performance using ISM approach. *Journal of Project Management*, 37–50.

Kumar, R., Kumar, V. and Singh, S. (2018). Work culture enablers: Hierarchical design for effectiveness and efficiency. *International Journal of Lean Enterprise Research*, 2(3), 189.

Kumar, R., Sindhwani, R., Arora, R. and Singh, P. L. (2021). Developing the structural model for barriers associated with CSR using ISM to help create brand image in the manufacturing industry. *International Journal of Advanced Operations Management*, 13(3), 312–330.

Lechner, C. and Gudmundsson, S. V. (2012). Entrepreneurial orientation, firm strategy and small firm performance. *International Small Business Journal: Researching Entrepreneurship*, 32(1), 36–60.

Link, P. (2016). How to Become a Lean Entrepreneur by Applying Lean Start-Up and Lean Canvas? *Innovation and Entrepreneurship in Education (Advances in Digital Education and Lifelong Learning*, Vol. 2), Emerald Group Publishing Limited, Bingley, 57–71.

Masudin, I., Sumah, B., Zulfikarijah, F. and Restuputri, D. P. (2021). Effect of information technology on warehousing and inventory management for competitive advantage. In *Handbook of research on innovation and development of E-commerce and E-business in ASEAN*. IGI Global, 570–593.

Maurya, A. (2012). *Running lean: Iterate from plan that works.* O'Reilly Media.

Maurya, A. (2018). *Scaling lean.* Penguin Putnam Inc.

McElwee, G. and Frith, K. (2008). The entrepreneurial wide boy. A modern morality tale. *International Journal of Entrepreneurship and Small Business*, 6(1), 80.

Merritt, T. (1974). Forecasting the future business environment—the state of the art. *Long Range Planning*, 7(3), 54–62.

Mittal, V. K., Sindhwani, R. and Kapur, P. K. (2016). Two-way assessment of barriers to lean—green manufacturing system: Insights from India. *International Journal of System Assurance Engineering and Management*, 7(4), 400–407.

Mittal, V. K., Sindhwani, R., Kalsariya, V., Salroo, F., Sangwan, K. S. and Singh, P. L. (2017). Adoption of integrated lean-green-Agile strategies for modern manufacturing systems. *Procedia CIRP*, 61, 463–468.

Müller, J. M., Buliga, O. and Voigt, K. I. (2018). Fortune favors the prepared: How SMEs approach business model innovations in Industry 4.0. *Technological Forecasting and Social Change*, 132, 2–17.

Naffziger, D. W., Hornsby, J. S. and Kuratko, D. F. (1994). A proposed research model of entrepreneurial motivation. *Entrepreneurship Theory and Practice*, 18(3), 29–42.

Nagar, D., Raghav, S., Bhardwaj, A., Kumar, R., Singh, P. L. and Sindhwani, R. (2021). Machine learning: Best way to sustain the supply chain in the era of industry 4.0. *Materials Today: Proceedings*, 47, 3676–3682.

Osterwalder, A. and Pigneur, Y. (2010). *Business model generation: A handbook for visionaries, game changers, and challengers* (Vol. 1). John Wiley & Sons.

Petru, N., Pavlák, M. and Polák, J. (2019). Factors impacting startup sustainability in the Czech Republic. *Innovative Marketing*, 15(3), 1–15.

Reichhart, A. and Holweg, M. (2007). Lean distribution: Concepts, contributions, conflicts. *International Journal of Production Research*, 45(16), 3699–3722.

Ries, E. (2011). *The lean startup: How today's entrepreneurs use continuous innovation to create radically successful businesses*. Crown Business.

Ries, E. (2017). *The startup way: How entrepreneurial management transforms culture and drives growth*. Penguin UK.

Robertson, M., Collins, A., Medeira, N. and Slater, J. (2003). Barriers to start-up and their effect on aspirant entrepreneurs. *Education and Training*, 45(6), 308–316.

Rosli, A. and Chang, J. (2020). *Team entrepreneurial learning: Building sustainable businesses. In How to become an entrepreneurship educator*. Edward Elgar Publishing, 145–152.

Roy, R. (2011). *Entrepreneurship*. Oxford University Press.

Safar, L., Sopko, J., Bednar, S. and Poklemba, R. (2018). Concept of SME business model for industry 4.0 environment. *Tem Journal*, 7(3), 626.

Santisteban, J., Mauricio, D. and Cachay, O. (2021). Critical success factors for technology-based startups. *International Journal of Entrepreneurship and Small Business*, 42(4), 397.

Shah, R. and Ward, P. T. (2003). Lean manufacturing: Context, practice bundles, and performance. *Journal of Operations Management*, 21(2), 129–149.

Shah, R. and Ward, P. T. (2007). Defining and developing measures of lean production. *Journal of Operations Management*, 25(4), 785–805.

Shahab, Y., Chengang, Y., Arbizu, A. D. and Haider, M. J. (2019). Entrepreneurial self-efficacy and intention: Do entrepreneurial creativity and education matter? *International Journal of Entrepreneurial Behavior & Research*, 25(2), 259–280.

Shanker, K., Shankar, R. and Sindhwani, R. (2019). Advances in industrial and production engineering. In *Select Proceedings of FLAME 2018 Book Series*. Springer-Nature.

Sindhwani, R., Afridi, S., Kumar, A., Banaitis, A., Luthra, S. and Singh, P. L. (2022b). Can industry 5.0 revolutionize the wave of resilience and social value creation? A multi-criteria framework to analyze enablers. *Technology in Society*, 101887.

Sindhwani, R., Gupta, R. D., Singh, P. L., Kaushik, V., Sharma, S., Phanden, R. K. and Kumar, R. (2021a). Identification of factors for lean and Agile manufacturing systems in rolling industry. In *Lecture notes in mechanical engineering.* Springer, Series, 367–378.

Sindhwani, R., Hasteer, N., Behl, A., Varshney, A., and Sharma, A. (2022a). Exploring "what," "why" and "how" of resilience in MSME sector: A m-TISM approach. *Benchmarking: An International Journal*, In press (ahead-of-print).

Sindhwani, R., Kumar, R., Behl, A., Singh, P. L., Kumar, A., and Gupta, T. (2021b). Modelling enablers of efficiency and sustainability of healthcare: A m-TISM approach. *Benchmarking: An International Journal*, 29(3), 767–792.

Singh, J., Gandhi, S. K. and Singh, H. (2019). Assessment of implementation of lean manufacturing in manufacturing unit-A case study. *International Journal of Business Excellence*, 1(1), 1.

Singh, P. L., Sindhwani, R., Sharma, B. P., Srivastava, P., Rajpoot, P. and Kumar, R. (2022). Analyse the critical success factor of green manufacturing for achieving sustainability in automotive sector. In *Recent trends in industrial and production engineering.* Springer, 79–94.

Smith, W. K., Binns, A. and Tushman, M. L. (2010). Complex business models: Managing strategic paradoxes simultaneously. *Long Range Planning*, 43(2–3), 448–461.

Spender, J., Corvello, V., Grimaldi, M. and Rippa, P. (2017). Startups and open innovation: A review of the literature. *European Journal of Innovation Management*, 20(1), 4–30.

Subrahmanya, M. H. (2021). *Entrepreneurial ecosystems for tech Start-ups in India: Evolution, structure and role.* Walter de Gruyter GmbH & Co KG.

Teece, D. J. (2010). Business models, business strategy and innovation. *Long Range Planning*, 43(2–3), 172–194.

Wahl, D. and Munch, J. (2021). Industry 4.0 entrepreneurship: Essential characteristics and necessary skills. IEEE International Conference on Engineering, Technology and Innovation (ICE/ITMC).

Walsh, B. (2009). Value is the core of your startup. In *The web startup success guide.* Apress, Springer, 25–47.

Wirtz, B. W., Pistoia, A., Ullrich, S. and Göttel, V. (2016). Business models: Origin, development and future research perspectives. *Long Range Planning*, 49(1), 36–54.

Womack, J. P., Jones, D. T. and Roos, D. (1990). *Machine that changed the world.* Simon & Schuster.

Yang, X., Sun, S. L. and Zhao, X. (2018). Search and execution: Examining the entrepreneurial cognitions behind the lean startup model. *Small Business Economics*, 52(3), 667–679.

Yavuztürk, H., Kalender, Z. T. and Vayvay, O. (2019). The role of universities in Industry 4.0 era: Entrepreneurship and innovation perspectives. In *Technological developments in industry 4.0 for business applications.* IGI Global, 50–70.

Zarezankova-Potevska, M. (2021). The Role of Education for Creation of Entrepreneurship Society. In: Ince-Yenilmez, M., Darici, B. (eds) *Engines of Economic Prosperity.* Palgrave Macmillan, Cham.

Zollo, M., Heimeriks, K., Mulotte, L., Ren, C. and Schijven, M. (2016). Beyond the experience curve: Learning and selection in corporate development activities. *Academy of Management Proceedings*, 2016(1), 16716.

2 Mapping Entrepreneurial Resilience Using Bibliometric Analysis— Future Agenda for Research in the Industry 4.0 Era

Shalini Rahul Tiwari and Shreya Homechaudhuri

CONTENTS

2.1 Introduction..21
2.2 Entrepreneurial Resilience: Concept and Beyond 22
    2.2.1 Evolution of the Entrepreneurial Resilience.......................................23
    2.2.2 Personal Predispositions Affecting Entrepreneurial Resilience..........23
    2.2.3 Contextual Studies on Entrepreneurial Resilience.............................24
    2.2.4 Resilience and Community Culture...26
    2.2.5 Resilience and Family Culture...26
2.3 Research Gaps 27
2.4 Research Questions..28
2.5 Research Methodology 28
2.6 Results..29
    2.6.1 Number of Publication..29
    2.6.2 Top Contributors 29
    2.6.3 Research Category 33
2.7 Analysis, Discussion and Future Research...35
References...35

2.1 INTRODUCTION

Forming new ventures is considered a fundamental requirement for sustained entrepreneurial growth, which is vital for a nation's long-term economic growth and development of a nation. In the era of Industry 4.0, entrepreneurs serve as the backbone of an economy, stimulating consumption by offering necessary and desirable goods and

DOI: 10.1201/9781003256663-2

services. Entrepreneurial activity in general, along with consumption and production cycles in particular, are often interspaced with unforeseen and unpredictable forces, such as natural disasters, economic crises, pandemics, etc. The impact of such events is usually long-lasting, especially on businesses, whereby some could also succumb to permanent closure. For instance, in 2008, the housing crisis and speculative investment banking practices led to the faltering and eventual crashing of the US economy, dragging the world economy down with it. The effects of the crash were realized across a vast horizon, spanning, for instance, the entire banking industry, while affecting in the process, people's lives, incomes, and livelihoods. The situation was further aggravated by the spill-over effect of the world economy, ushering in a harmful cycle of a credit crunch, the immediate consequences of which were felt by smaller businesses. They were unable to take loans/credit, so they were compelled to tighten their production cycles, affecting supplies while also affecting their ability to retain critical human resources. Drawing from this example, several pertinent questions arise- how do entrepreneurs survive such times? Do all businesses close down during such challenging times?

It has been observed that some business leaders and entrepreneurs often bounce back post facing initial setbacks. Thus, another pertinent question arises: what effectively drives entrepreneurs to display this sort of resilience? In order to address the same, it is imperative to understand the role of "resilience" in promoting entrepreneurial growth. As a concept, "resilience" is more related to Social Science, whereby it has a dualistic nature, one in which, it is perceived as an ability to "bounce back." On the other hand, it is based on a more dynamic consideration of the adaptive capacity of an individual as a social being. The current scenario of the ongoing Covid-19 pandemic has given us an opportunity to capture the concept of "resilience" in greater detail (Bacq et al., 2020). This study aims to review extant literature, specifically in the field of Entrepreneurial Resilience (ER). Specifically, our objective is to draw significant insights culled from literature, and thereby attempt to frame the future research agenda on this fascinating topic. The remainder of the paper is as follows: the next section elaborates upon the concept of "resilience," while tracing the evolution of the construct, the factors or variables that have influenced ER, along with the moot research questions of the study. The second section describes the research methodology, while the last section analyzes the results and discusses ideas for further ideas for research.

2.2   ENTREPRENEURIAL RESILIENCE: CONCEPT AND BEYOND

As we have seen earlier, resilience has been explained in many ways by different scholars in the past. For instance, according to Masten (2001), resilience has been defined as "excellent results notwithstanding substantial risks to adaptation or development." Richardson (2002) defined it as "the process of coping with adversity, change, or opportunity in a way that results in the identification, fortification, and enrichment of resilient traits or protective elements." According to Bonanno (2004), resilience represents the capacity to preserve both physical and psychological security in the face of potentially disrupting events. Interestingly, despite the varied differences, definitions of resilience do have some parallels too, such as adjusting to changing conditions, overcoming adversity, and rebounding while making ongoing improvements. In terms of the working definition of resilience, possibly the best suited in terms of incorporating all aspects of today's changing world could be Bonanno's (2004) version—the ability to bounce back from adversity and cope with traumatic events (Zautra et al., 2010), as well as confront problems successfully and create positive outcomes despite adversity, is referred to as resilience (Werner, 1989; Luthar et al., 2000; Masten, 2001).

2.2.1 EVOLUTION OF THE ENTREPRENEURIAL RESILIENCE

For better understanding and insights into Resilience as a concept, it is vital to study its evolution. The term's first application may be traced back to the field of Engineering and Ecology. Over the years, economic conditions soon made way for the emergence of "adaptive resilience" as a concept, signifying the ability to withstand market and environmental shocks. It may be interesting to note here that earlier research on "resilience" apart from the topic of engineering and ecology, also focused on disquieting events experienced by children (e.g., chronic diseases, early parent bereavement, etc.) (Rutter, 1985; Garmezy, 1991; Werner, 1995; Masten and Garmezy, 1985). In fact, these studies eventually paved the way for additional studies that chose to focus on the psychopathological aspects of Resilience. Several of these studies concluded: "resilience appears to be a common phenomenon arising from ordinary human adaptive processes." Thus, over time, the focus of studies on Resilience evolved, whereby the research domain broadened from studying children to looking at adults not only from the perspective of psychopathology but even other subject areas. For instance, Bonanno (2004) expanded Masten's (Masten and Garmezy, 1985) concept of Resilience, covered in psychopathology, to encompass potentially traumatic circumstances that individuals under typical situations may probably encounter at least once in their lives. The findings of these studies have effectively laid the foundation for further research to explore the extent of individual Resilience, stressing upon the significance of identifying the significant aspects that go on to determine it in times of any crisis. Herein, it may also be pertinent to note that crisis-related studies in entrepreneurship literature had increased considerably in the last decade (Doern, 2016).

2.2.2 PERSONAL PREDISPOSITIONS AFFECTING ENTREPRENEURIAL RESILIENCE

A person's Resilience vis a vis Entrepreneurial self-efficacy (ESE) are interrelated, and thereby form two important elements of entrepreneurial thinking, which sheds light on an entrepreneur's intentions across varied contexts. Due to its relevance to entrepreneurship, "resilience" has drawn considerable attention of business scholars in the recent past (Bullough and Renko, 2014; Corner et al., 2017; Shepherd et al., 2020). In fact, research on Entrepreneurial Resilience (ER) today has been acknowledged as a vital component in understanding entrepreneurial behavior, especially in adapting and overcoming uncertainties and challenges, while continually learning from previous failures. Several studies have compared resilient entrepreneurs as opposed to non-resilient ones. They have weighed in on the possibility of resilient entrepreneurs being more successful in business than non-resilient ones (Ayala and Manzano, 2014); or how resilient entrepreneurs adapt to changes while rebounding in challenging scenarios (Bullough et al., 2013). Going by these studies, it may thereby be inferred that the element of "self-efficacy" among entrepreneurs is intrinsically linked to ER (Bullough et al., 2014).

Additionally, it may also be noted that apart from self-efficacy, another vital component of ER includes an individual's locus of control, which in turn, may be linked to the Attribution theory, propounded by Heider (1958). The theory examines how people tend to use attributions in order to determine the cause-and-effect of relationships. It has three aspects: the locus of causality, stability, and controllability. In summary, this theory essentially describes how entrepreneurs ascribe their success or failure (Weiner, 1986). Almost extending Heider's (1958) theory, Rotter (1966) stated that an individual's perception of the consequence of a certain incident is either within or beyond his/her control. In the entrepreneurial context, an entrepreneur is assumed to have a robust internal locus of control, indicating they firmly believe in their ability to influence outcomes with their efforts, capabilities, and/or talents. That is why we see most entrepreneurs having a tendency to take charge of their own destiny, irrespective of others' opinions, especially in their business matters (McClelland, 1961). Such differential entrepreneurial thinking must be channelized into a more researched and strategized entrepreneurial process in order to sustain long term in business or when faced with adversities, thus laying the platform for further studies on determinants of individual personality components in line with the qualities of entrepreneurship and Resilience.

2.2.3 CONTEXTUAL STUDIES ON ENTREPRENEURIAL RESILIENCE

The concept of Resilience has also been explored widely even in other contexts, which effectively has resulted in widening the concept. For instance, it has been investigated in terms of tourism-related risks (Ngoasong et al., 2016), terrorism (Branzei and Abdelnour, 2010), conflict situations and national security crises (e.g., countries facing wars), external turbulences (e.g., recessions or financial crises), natural disasters (e.g., hurricanes, floods, earthquakes, tsunamis), human-induced crises accidents (Doern, 2016). Contextual studies have also taken into account a plethora of variables that significantly impact entrepreneurship to predict the determining factors of resilience in case of any economic downturn. Promoting entrepreneurial spirit resulted in enhanced entrepreneurship that positively influences important macroeconomic parameters, such as employment, innovation, and overall sustained development. Innovative entrepreneurial ventures and activities add value to improving a nation's economic conditions by creating jobs, thereby becoming a precedence for several regimes worldwide (Kraus et al., 2019). Thus, even though entrepreneurs are the first-level commercial entities most significantly affected by any kind of crisis, they are also the primary agents for resurgence and subsequent economic development because of their role and implication as a vehicle of modernization and creation of jobs (Voda et al., 2019). Therefore, studies on how entrepreneurs handle and overcome these adversities, especially in the case of small and medium enterprises, become highly relevant in gaining a better understanding of crises management while limiting the impact (Cucculelli and Bettinelli, 2015). In fact, entrepreneurs' innovative pattern effectively triggers the development of services and products based upon customers' needs.

Moreover, this innovation entices entrepreneurs to update their existing business models regarding how their companies function to exploit and reap benefits from new opportunities and create value while practicing Resilience (George and Bock, 2011). Thus, it is commonly believed that the seed of entrepreneurship should effectively be sown right at the level of primary education, whereby students must be taught to develop entrepreneurial skills, including the aspect of Resilience. Importantly, this education should look to enrich the students' skills to better adapt to changing business environments, according to the different entrepreneurial cultures, values, and aspirations (Dana, 2001; Dana and Dana, 2005). The European Commission (EC) in this regard, stated that it is vital to build entrepreneurial capabilities that enable people to develop qualities like Resilience, so that existing companies have a "ready" employees, equipped with the abilities required to adapt to the different working milieu (European Commission, 2020). However, Duchek (2018) noted that despite the important implications, that entrepreneurs have for innovation and economic development of a nation per se, very few studies have actually covered the issue of ER, along with the factors that go on to strengthen it in individuals/entrepreneurs. The author, in fact, reviewed several biographies of successful entrepreneurs and attempted to identify and segregate factors and relevant attitudes that underlie the resilience capability of entrepreneurs. By and large, the findings of these studies emphasize certain aspects of entrepreneurship that are considered vital, especially in terms of governance of emerging economies (e.g., India). Importantly, such economies are prone to ambiguity, and hence do have a need for planned guidance, so that they can face various economic turmoil, both within their own land, and also across geographies, and thereby are more prepared to both adopt and adapt to the recovery process in case of business/economic failures.

Thus, entrepreneurship is critical for local, regional, and national development (Apostolopoulos et al., 2020). The entrepreneurial characteristics, such as Resilience, motivation, intention, etc. must be imparted through formal education and inculcated in citizens to understand their environment better (Nyadu-Addo and Mensah, 2018; Beltran Hernandez de Galindo et al., 2019). This serves as an inspiration for the universities worldwide to take this as an opportunity to craft scopes to accomplish their objective of teaching entrepreneurship-related courses to kindle Resilience and generate new-age jobs (Maritz et al., 2020). Crises experienced because of difficult times such as—COVID-19, financial meltdown, tsunamis, etc. have led entrepreneurs to confront challenges, constraints and threats (Doern et al., 2019). A comprehension of Resilience as a concept is essential, as it ensures that an organization continues functioning reliably during such crises periods (Williams et al., 2017). Among the studies conducted in this domain, the most significant volume of work is associated with the period before the events caused the crises. Notably, only a few papers talked about how individuals respond to crises. These studies essentially recommended that more research be done to determine what entrepreneurs do in response to problems and how they could improve Resilience in their communities (Korber and McNaughton, 2017).

Interestingly, an entrepreneurial start-up's failure is often considered a social stigma in some societies (Klimas et al., 2021). As a result, societal norms affect entrepreneurial activity and Resilience within a particular society at times. For example, in the United States, failure "generally" is more accepted and is seen as a learning experience. However, in Japan and Europe on the other hand, failure is often perceived as a social disgrace (European Commission, 2003). Such disparities and inequalities between countries, regions, and cultures allude to the fact that resilient entrepreneurial behavior is more readily accepted and adopted in societies where a failure of an entrepreneurial endeavor is not perceived as a societal disgrace (Vaillant et al., 2007). It may be pertinent to note that entrepreneurial activity itself varies significantly among countries, regions, and cultures. These variances, in turn, could be linked to the societal shame associated with failure. Thus, with the emergence of "culture" as a significant impacting factor in ER studies, the past decade has seen an increasing number of researches being done to explore the "cultural" aspect's contribution to ER.

2.2.4 Resilience and Community Culture

The ability of businesses to design and adapt to the outcomes of natural disasters on indigenous societies can be critical in determining how they actually respond to those disasters. This is particularly relevant for Scotland and the UK, where there is a robust legislative focus on sanctioning local companies to become "irrepressible or resilient" to extreme weather such as heavy rains or snow, and/or excessive heat (Cabinet Office, 2011; Preparing Scotland Report, 2013). Further, "Preparing Scotland" had issued an advisory to businesses for being resilient (2016) "Developing and maintaining business resilience within an organization is likely to create chances to promote resilience externally," As a result, the policy focus on strengthening small companies is primarily motivated by the notion that increasing company resilience would effectively reduce risk, and encourage resilient behavior in other areas, contributing eventually to a comprehensive communal resilience. Along similar lines, Paton et al. (2017) suggested that enterprises having the ability to recuperate quickly can carry on to supply jobs and services, especially within afflicted regions. Due to this, several scholars have proposed that both company resilience and community resilience are intertwined (Huggins et al.,2014). However, there has been no empirical research that has explored this relationship until now (Norris et al., 2008).

2.2.5 Resilience and Family Culture

Sustainable Family Business Theory (SFBT) has been a predominant theory grounded on the concept of Resilience, pertaining primarily to family-owned businesses. SFBT suggests that family incorporates capital stocks and resources in family and family-owned firms, while interpersonal processes ensure family business sustainability (Danes and Brewton, 2012). In other words, SFBT has primarily focused on creating and preserving family resilience because relational family processes serve effectively as the context that endures entrepreneurship (Eddleston and

Morgan, 2014). Furthermore, relational ethics is believed to be built on a feeling of justice and/or equality in relationships. Boszormenyi-Nagy and Krasner (1986) stated that when people are both compelled to give to the network and entitled to receive from it, the balance (or imbalance) of interpersonal interactions in familial connections is registered in the family ledger. In fact, these very ideas go on to serve as a framework for interpreting family actions, which in turn, are symbolic of core family processes and procedures. On the other hand, relational transaction configurations (family standard operating procedures in times of stability) involve processes that are mostly unseen patterns generated, sustained, and improved by a family. This, in effect, enables the family to be resilient to change and/or hardships while attempting to build some form of social capital for future use. However, there are two crucial points to remember—one must first ascertain apparent conducts that indicate fundamental family practices of interest. Secondly, once the observable behaviors have been assessed, the results must be integrated to determine the quality and quantity of the process. The observable actions of functionally robust families, which address the initial significant point of assessing relational practices, are the concepts that make up relational ethics (Hanson, 2019) that enforce Resilience among family-owned businesses.

Legacy is a dynamic multigenerational concept that encompasses the past and the present and is sometimes misinterpreted as the family history changes over a period of time. Individuals hold indiscernible allegiances across generations, even when creating a different family entity; legacy is passed down through generations (Boszormenyi-Nagy and Krasner, 1986). When one generation infuses an entrepreneurial essence in the next, the next generation's management skills become inclined towards keeping in line with the generational Culture of entrepreneurship (Jaskiewicz et al., 2015). Thus, it is generally believed that family-owned businesses are more resilient than professionally managed businesses.

2.3 RESEARCH GAPS

Based on our analysis of extant literature, we identified three reservations about existing methods to entrepreneurial Resilience (ER) research. First, extant research on ER didn't consider the distinctions between observable contexts and psychopathological disorders (e.g., Connor—Davidson resilience scale). Specifically, researchers studied the concept of Resilience in children in the entrepreneurial context or used tools established to measure it to investigate the attitude and performance of resilient entrepreneurs (Ayala et al., 2014; Bullough et al., 2014; Hedner et al., 2011). So, only some of the items have been selectively helpful in effectively capturing the qualities of ER. The next point of concern includes the absence of a comprehensive view of ER under different economic structures. Notably, studies thus far have primarily concentrated on the psychological aspects of ER (Ayala et al., 2014; Bullough et al., 2013; Bullough et al., 2014); at most, they have equated similar concepts (e.g., confidence). Thirdly, researchers in the past have been unable to get a thorough understanding of ER due to this limited perspective. It may be noted herein that the features of "resilience" per se, may vary depending upon the situation. Importantly, as multiple cognitive, social, emotional, and financial factors do influence ER, there should be additional variables that should be drawn in order to understand the process of ER holistically. Thus, domain-specificity should imperatively be included in ER research (Luthar et al., 2000; Tusaie et al., 2004).

Our review of literature traces the evolution of the domain of ER over time. We tried to identify a few significant themes that are of research interest to academicians worldwide. In the process, we observed that even though there are significant theories that support the psychological, economic and social aspects of Resilience, there has clearly been a gap in domain-specific theories and measures. The reason behind such a gap could be because the topic is at an elementary stage of development, having gained significance only over the past decade, with frequent turbulences faced by global economies. Without any domain-specific guiding theory for ER, measuring the progress made toward theory building, and enhancing upon ER development practices, does become a challenge. Nevertheless, one must acknowledge the work done by extant literature, which has identified some of the important determining factors that need to be further explored and developed in resilience studies (Bullough et al., 2014). One such important determinant is culture; the cultural environment persisting in an organization or the cultural values embedded within individuals forms the base for the adaptive capacity during challenging times. Furthermore, due to the novelty of the domain, cultural studies with regard to ER have been in a formative stage, and there's a wide scope for further studies, encompassing this aspect. From existing literature, we also understand the importance of domain-specific contextual studies; and we believe that further studies should seek to explore the efficacy of ER, when it comes to emerging economies like India. Notably, these economies require strategic planning and intensive development, which could transform them into a booming economy.

Based on the discussion above, the following research questions emerge:

2.4   RESEARCH QUESTIONS

1. What is the nature of research done on ER in the last few years?
2. What elements of ER have been studied to strengthen the construct, context, and process?
3. Which research clusters have emerged in the ER domain across the world?

2.5   RESEARCH METHODOLOGY

For data acquisition, we referred to Scopus, wherein we used the following keywords' combinations in order to generate an extensive study sample in the field of ER. They include "entrepreneurial resilience," "resilient entrepreneurship," and "business resilience." The search period covered two decades, starting from 2002 until 2021. Once we completed the search, we scanned all the records manually with an aim to eliminate irrelevant data. Our final data included 159 publication records across various journals after this activity. Herein, it is essential to understand that our field of study on ER is a relatively emerging topic. Therefore, the timeframe of our study is completely inclusive of all the work that has been done to date across the globe.

Extracting a bibliometric web from the social network is a practical technique for analyzing existing research. This method effectively enhances the systematic thinking of researchers by mapping networks (Pauna et al., 2019). For network analysis, we used VOS-viewer (version 1.6.13). Specifically, this software can cluster, map networks, and visualize bibliometric information (Van Eck et al., 2010).

We analyzed data by looking at several categories, such as the leading contributors in the field of ER, vis a vis their linkage networks with other authors working in the same field; collaboration network of countries; network of citations for various countries in order to show how researchers actually refer to other countries' scientific studies. Thus, we adopted the following process after identifying some of the top contributors, along with their relations and interrelations. In the following step, we classified the category of research. Keywords were used to select the article. Since the focus of the research is to map the ER domain, keywords such as Entrepreneurial Resilience, Resilient Entrepreneurship, Resilience were used along with the keywords network for analyzing the research content.

2.6 RESULTS

The bibliometric analysis has been presented under the following categories:

2.6.1 NUMBER OF PUBLICATIONS

To understand the research patterns, trend analysis was done to identify the number of publications per year on ER across all the journals. The results are shown in Table 2.1.

Further, Figure 2.1 shows the growing importance of research on ER over the years. There has been a steep increase in research publications from 2019 to 2021, indicating thereby that more relevant research themes in this field have emerged in recent times. Additionally, there has been a substantial increase in the number of publications from 2015 onwards. However, the growth rate was not steady till 2019, after which, it picked steam. Notably, 2021 alone accounted for 26% of all the publications that have been done in the domain of ER.

2.6.2 TOP CONTRIBUTORS

To understand the extent and reach of the domain across the globe, it was important to identify the top contributors who have been studying and researching on resilience over the years. Figure 2.2 depicts the top author network collaborations that have significantly contributed to research on ER. Table 2.2 shows that the author Lee J. has the highest number of publications in this domain, along with a considerably high number of citations for research in this field. Figure 2.3 shows the clusters of the top authors according to their collaboration linkages.

Figure 2.4 demonstrates the top countries' authorship networks with each other. Table 2.3 shows that the United Kingdom has the highest number of author

TABLE 2.1

Number of Publications per Year and Contribution to Total Research

Year	Number of Publications	Percentage of total research (%)
2021	42	26
2020	27	17
2019	11	7
2018	14	9
2017	16	10
2016	12	8
2015	8	5
2014	3	2
2013	5	3
2012	4	3
2011	6	4
2010	1	1
2009	1	1
2008	2	1
2007	2	1
2006	4	3
2005	1	1
Total	159	

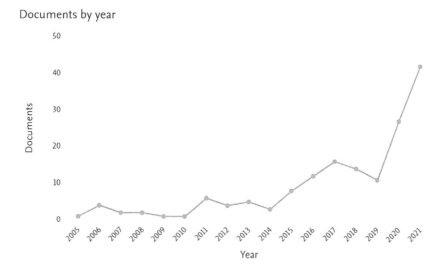

FIGURE 2.1 Number of publications per year.

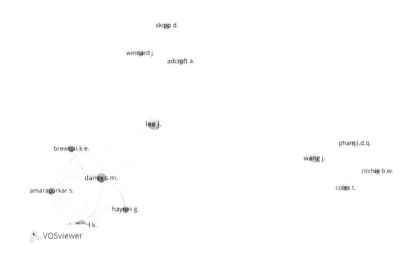

FIGURE 2.2   Top author networks.

TABLE 2.2
Top Authors and Their Link Strengths

Author	Documents	Citations	Total link ⌄ strength
lee j.	3	86	9
adnan a.h.m.	2	8	7
jaafar r.e.	2	8	7
mohtar n.m.	2	8	7
nasir z.a.	2	8	7
danes s.m.	2	69	6
stafford k.	2	69	6
amarapurkar s.	1	57	5
brewton k.e.	1	57	5
de lange w.	1	5	5
de wet b.	1	5	5
haynes g.	1	57	5
haywood l.	1	5	5
marshall m.i.	2	26	5
musvoto c.	1	5	5
stafford w.	1	5	5
watson i.	1	5	5
adiguna r.	1	0	4
agarwal s.	1	2	4

Cluster 1 (5 items)
amarapurkar s.
brewton k.e.
danes s.m.
haynes g.
stafford k.
Cluster 2 (4 items)
coles t.
pham l.d.q.
ritchie b.w.
wang j.
Cluster 3 (4 items)
adcroft a.
lee j.
skipp d.
winnard j.

FIGURE 2.3   Clusters of top authors.

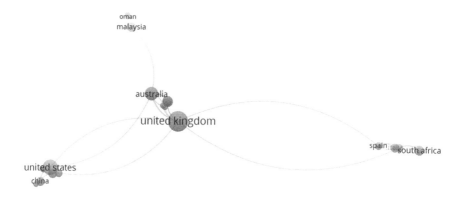

FIGURE 2.4   Top countries authorship network.

collaborations in the ER domain, followed by the United States and Australia. Importantly, ER has comparatively been under-researched in India. India has about seven publications on the topic, with just three collaborations with Canada and North Macedonia (Figure 2.5). The larger global network included five clusters; the first cluster contained two major contributors from the United Kingdom and Australia, along with Indonesia, New Zealand, Nigeria, and Singapore. The second cooperation cluster was between Italy, Netherlands, Norway, Poland, South Africa, and Spain. The third cluster contained Canada, China, India, and Taiwan. Notably, the fourth and fifth clusters included just three-member clusters, namely, Malaysia, Oman, and Pakistan, and Columbia, France, and the United States.

TABLE 2.3

Top Countries Authorship Networks

Country	Documents	Citations	Total link strength ∨
united kingdom	31	243	12
australia	14	303	10
netherlands	5	41	7
united states	18	227	7
spain	5	7	6
canada	5	23	4
greece	1	3	4
italy	6	38	4
new zealand	9	219	4
poland	3	6	4
china	6	12	3
india	7	26	3
indonesia	8	113	3
nigeria	3	2	3
taiwan	3	12	3
chile	2	5	2
france	4	70	2
lithuania	3	5	2
malaysia	7	17	2

Cluster 6 (3 items)

canada

india

north macedonia

FIGURE 2.5   Cluster 6.

Figure 2.6 shows the author citations from different nations, collaborating in the publication domain. Notably, the citation network of these top authors from various nations shows eight clusters, showing how the authors cited works from other countries and referred to each other's publications.

2.6.3   RESEARCH CATEGORY

Keywords represent the articles' core content, and they showed the research area that is being explored in existing studies. On the other hand, VOS-viewer is a powerful software, identifying the top keywords, while visualizing the relations between these

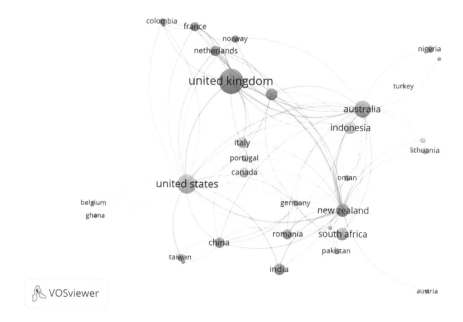

FIGURE 2.6   Citation network of top countries.

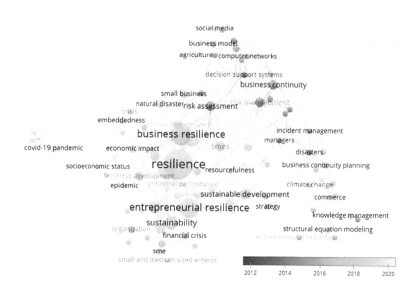

FIGURE 2.7   Top keyword clusters.

keywords. Notably, the recorded results in Scopus showed the relevant keywords for the given timeframe. Figure 2.7 shows the top keywords within the network, along with their clusters. Furthermore, from the network map, we note that till early 2018, the concept of Resilience, mainly revolved around the sub-concepts of disaster management, business model planning, knowledge management, and entailed very little scope of study with respect to ER. Additionally, the concepts of ER, business resilience and sustainable development have started emerging since 2019. Importantly herein, the ongoing Covid-19 pandemic may have triggered the shifting focus on ER.

2.7   ANALYSIS, DISCUSSION, AND FUTURE RESEARCH

This bibliometric analysis on ER does give us a clear understanding of where the domain lies, the extent of research in the area, along with the research gaps, which in turn, provides scope for further research. Moreover, several interesting and important insights have emerged from this analysis; first, we observed that studies on Resilience peaked around times when the world was faced with certain economically challenging scenarios, like the US housing crisis, the OPEC oil price crisis or a pandemic like Covid-19. All these crises created a situation for businesses and entrepreneurs to exhibit Resilience, by taking critical survival decisions. Secondly, from our literature review, we observed that culture does play a critical role in studies on Resilience. Specifically, it helps shape entrepreneurial capability to handle crises, and build a brand personality for organizations. Culture also instills motivation in an entrepreneur to be adaptive, and thereby be able to make innovative changes in the face of challenging situations. Interestingly, we have not found enough studies on culture from our bibliometric analysis, which in effect, points towards a potential research gap. Thirdly, studies on resilience research have been concentrated in developed economies like the UK and the US, with very few contextual studies on developing and less developing economies. These economies are in dire need of research on ER, as they face considerable macroeconomic risks that go on to affect the business environment. Thus, it becomes imminent for researchers to delve deep into the contexts of the emerging market to understand the dimensions of the construct, impact, outcomes, and so on. Fourthly, no substantial studies have been conducted to understand the theoretical foundations of ER. In order to develop a richer understanding of the ER process, ER needs to be studied from several theoretical lenses. Lastly, it is interesting to note that the keywords "business" and "resilience" have not shown enough research articles. Does it imply that Resilience is an individual-level construct? Or has it not been researched by the authors? Or has it been studied with some other construct? These questions need to be explored thoroughly. Our study provides enough rationale for researchers to deep dive into this area.

REFERENCES

Apostolopoulos, N., Ratten, V., Stavroyiannis, S., Makris, I., Apostolopoulos, S. and Liargovas, P. (2020). Rural health enterprises in the EU context: A systematic literature review and research agenda. *Journal of Enterprising Communities: People and Places in the Global Economy*, In press (ahead-of-print). 10.1108/JEC-04-2020-0070.

Ayala, J. C. and Manzano, G. (2014). The resilience of the entrepreneur. Influence on the success of the business. A longitudinal analysis. *Journal of Economic Psychology*, 42, 126–135.

Bacq, S., Geoghegan, W., Josefy, M., Stevenson, R. and Williams, T. A. (2020). The COVID-19 Virtual Idea Blitz: Marshaling social entrepreneurship to rapidly respond to urgent grand challenges. *Business Horizons*, 63(6), 705–723.

Beltrán Hernández de Galindo, M. D. J., Romero-Rodríguez, L. M. and Ramirez Montoya, M. S. (2019). Entrepreneurship competencies in energy sustainability MOOCs. *Journal of Entrepreneurship in Emerging Economies*, 11(4), 598–616.

Bonanno, G. A. (2004). Loss, trauma, and human Resilience: Have we underestimated the human capacity to thrive after extremely aversive events? *American psychologist*, 59(1), 20.

Boszormenyi-Nagy, I. and Krasner, B. R. (1986). *Between give and take: A clinical guide to contextual therapy*. Brunner/Mazel.

Branzei, O. and Abdelnour, S. (2010). Another day, another dollar: Enterprise resilience under terrorism in developing countries. *Journal of International Business Studies*, 41(5), 804–825.

Bullough, A. and Renko, M. (2013). Entrepreneurial Resilience during challenging times. *Business Horizons*, 56(3), 343–350.

Bullough, A., Renko, M. and Myatt, T. (2014). Danger zone entrepreneurs: The importance of Resilience and self—efficacy for entrepreneurial intentions. *Entrepreneurship Theory and Practice*, 38(3), 473–499.

Cabinet Office (2011). Annual report and accounts. https://assets.publishing.service.gov.uk/government/uploads/system/uploads/attachment_data/file/325326/41432_HC_Cabinet_Office_annual_report_2013_to_2014_print_ready.pdf

Corner, P. D., Singh, S. and Pavlovich, K. (2017). Entrepreneurial resilience and venture failure. *International Small Business Journal*, 35(6), 687–708.

Cucculelli, M. and Bettinelli, C. (2015). Business models, intangibles and firm performance: Evidence on corporate entrepreneurship from Italian manufacturing SMEs. *Small Business Economics,* 45(2), 329–350.

Dana, L. P. (2001). The education and training of entrepreneurs in Asia. *Education+ Training*, 43(8/9), 405–416.

Dana, L. P. and Dana, T. E. (2005). Expanding the scope of methodologies used in entrepreneurship research. *International Journal of Entrepreneurship and Small Business*, 2(1), 79–88.

Danes, S. M. and Brewton, K. E. (2012). Follow the capital: Benefits of tracking family capital across family and business systems. In *Understanding family businesses*. Springer, 227–250.

Doern, R. (2016). Entrepreneurship and crisis management: The experiences of small businesses during the London 2011 riots. *International Small Business Journal*, 34(3), 276–302.

Doern, R., Williams, N. and Vorley, T. (2019). Special issue on entrepreneurship and crises: business as usual? An introduction and review of the literature. *Entrepreneurship and Regional Development*, 31(5–6), 400–412.

Duchek, S. (2018). Entrepreneurial Resilience: a biographical analysis of successful entrepreneurs. *International Entrepreneurship and Management Journal*, 14(2), 429–455.

Eddleston, K. A. and Morgan, R. M. (2014). Trust, commitment and relationships in family business: Challenging conventional wisdom. *Journal of Family Business Strategy*, 5(3), 213–216.

European Commission (2003). Green paper on entrepreneurship in Europe, Commission of European Communities, Brussels. https://ec.europa.eu/invest-in-research/pdf/download_en/entrepreneurship_europe.pdf

European Commission (2020). Internal market, industry, entrepreneurship and SMEs. https://ec.europa.eu/growth/sectors/tourism/funding-guide/recovery-and-resilience-facility_en

Garmezy, N. (1991). Resiliency and vulnerability to adverse developmental outcomes associated with poverty. *American Behavioral Scientist*, 34(4), 416–430.

George, G. and Bock, A. J. (2011). The business model in practice and its implications for entrepreneurship research. *Entrepreneurship Theory and Practice*, 35(1), 83–111.

Hanson, A. H. (2019). *Public enterprise and economic development.* Routledge.

Hedner, T., Abouzeedan, A. and Klofsten, M. (2011). Entrepreneurial resilience. *Annals of Innovation & Entrepreneurship*, 2(1), 7986.

Heider, F. (1958). The naive analysis of action. In F. Heider (Ed.), *The psychology of interpersonal relations.* John Wiley & Sons Inc., 79–124. https://doi.org/10.1037/10628-004

Huggins, R. and Thompson, P. (2014). Culture, entrepreneurship and uneven development: A spatial analysis. *Entrepreneurship and Regional Development*, 26(9–10), 726–752.

Jaskiewicz, P., Combs, J. G. and Rau, S. B. (2015). Entrepreneurial legacy: Toward a theory of how some family firms nurture transgenerational entrepreneurship. *Journal of Business Venturing*, 30(1), 29–49.

Klimas, P., Czakon, W., Kraus, S., Kailer, N. and Maalaoui, A. (2021). Entrepreneurial failure: A synthesis and conceptual framework of its effects. *European Management Review*, 18(1), 167–182.

Korber, S. and McNaughton, R. (2017). Resilience and entrepreneurship: A systematic literature review. *International Journal of Entrepreneurial Behavior & Research*, 24. 10.1108/IJEBR-10-2016-0356

Kraus, S., Palmer, C., Kailer, N., Kallinger, F. and Spitzer, J. (2019). Digital entrepreneurship: A research agenda on new business models for the twenty-first century. *International Journal of Entrepreneurial Behavior and Research*, 25(2), 353–375. https://doi.org/10.1108/IJEBR-06-2018-0425

Luthar, S. S. and Cicchetti, D. (2000). The construct of Resilience: Implications for interventions and social policies. *Development and Psychopathology*, 12(4), 857–885.

Maritz, A., Perenyi, A., De Waal, G. and Buck, C. (2020). Entrepreneurship as the unsung hero during the current COVID-19 economic crisis: Australian perspectives. *Sustainability*, 12(11), 4612.

Masten, A. S. (2001). Ordinary magic: Resilience processes in development. *American Psychologist*, 56(3), 227.

Masten, A. S. and Garmezy, N. (1985). Risk, vulnerability, and protective factors in developmental psychopathology. In B. B. Lahey and A. E. Kazdin (Eds.), *Advances in clinical child psychology* (Vol. 8). Plenum Press, 1–52.

McClelland, D. C. (1961). Entrepreneurial behavior. In D. C. McClelland (Ed.), *The achieving society.* D Van Nostrand Company, 205–258. https://doi.org/10.1037/14359-006

Ngoasong, M. Z. and Kimbu, A. N. (2016). Informal microfinance institutions and development-led tourism entrepreneurship. *Tourism Management*, 52, 430–439.

Norris, F. H., Stevens, S. P., Pfefferbaum, B., Wyche, K. F. and Pfefferbaum, R. L. (2008). Community resilience as a metaphor, theory, set of capacities, and strategy for disaster readiness. *American Journal of Community Psychology*, 41, 12

Nyadu-Addo, R. and Mensah, M. S. (2018). Entrepreneurship education in Ghana – the case of the KNUST entrepreneurship clinic. *Journal of Small Business and Enterprise Development*, 25, 573–590.

Paton, D. and Johnston, D. (2017). *Disaster resilience: An integrated approach.* Charles C Thomas Publisher.

Pauna, V. H., Buonocore, E., Renzi, M., Russo, G. F. and Franzese, P. P. (2019). The issue of microplastics in marine ecosystems: A bibliometric network analysis. *Marine Pollution Bulletin*, 149, 110612.

Preparing Scotland Report (2013). Preparing Scotland: Business resilience guide. https://www.gov.scot/publications/preparing-scotland-having-promoting-business-resilience/pages/5/

Richardson, G. E. (2002). The metatheory of resilience and resiliency. *Journal of Clinical Psychology*, 58(3), 307–321.

Rotter, J. B. (1966). Generalized expectancies for internal versus external control of reinforcement. *Psychological Monographs: General and Applied*, 80(1), 1.

Rutter, M. (1985). Resilience in the face of adversity: Protective factors and resistance to psychiatric disorder. *The British Journal of Psychiatry*, 147(6), 598–611.

Shepherd, D. A., Saade, F. P. and Wincent, J. (2020). How to circumvent adversity? Refugee-entrepreneurs' Resilience in the face of substantial and persistent adversity. *Journal of Business Venturing*, 35(4), 105940.

Tusaie, K. and Dyer, J. (2004). Resilience: A historical review of the construct. *Holistic Nursing Practice*, 18(1), 3–10.

Vaillant, Y. and Lafuente, E. (2007). Do different institutional frameworks condition the influence of local fear of failure and entrepreneurial examples over entrepreneurial activity? *Entrepreneurship and Regional Development*, 19(4), 313–337.

Van Eck, N. J. and Waltman, L. (2010). Software survey: VOSviewer, a computer program for bibliometric mapping. *Scientometrics*, 84(2), 523–538.

Vodă, A. I. and Florea, N. (2019). Impact of personality traits and entrepreneurship education on entrepreneurial intentions of business and engineering students. *Sustainability*, 11(4), 1192.

Weiner, B. (1986). Attribution, emotion, and action. In R. M. Sorrentino and E. T. Higgins (Eds.), *Handbook of motivation and cognition: Foundations of social behavior*. Guilford Press, 281–312.

Werner, E. E. (1989). High-risk children in young adulthood: A longitudinal study from birth to 32 years. *American Journal of Orthopsychiatry*, 59(1), 72–81.

Werner, E. E. (1995). Resilience in development. *Current Directions in Psychological Science*, 4(3), 81–84.

Williams, T. A., Gruber, D. A., Sutcliffe, K. M., Shepherd, D. A. and Zhao, E. Y. (2017). Organizational response to adversity: Fusing crisis management and resilience research streams. *Academy of Management Annals*, 11, 1–70.

Zautra, A. J., Hall, J. S. and Murray, K. E. (2010). Resilience: A new definition of health for people and communities. In J. W. Reich, A. J. Zautra and J. S. Hall (Eds.), *Handbook of adult resilience*. The Guilford Press, 3–29.

3 Trends in Social Entrepreneurship Landscape of India
Past Contributions and Future Opportunities

Vikram Bansal and Deepthi B.

CONTENTS

3.1 Introduction... 39
3.2 What Distinguishes Social Entrepreneurship from Traditional
    Entrepreneurship?... 41
3.3 Attributes of Social Entrepreneurs.. 43
3.4 The Social Entrepreneurship Landscape in India 45
3.5 Challenges Faced by Social Entrepreneurs in India 46
3.6 Past Contributions of Social Entrepreneurs in India......................... 47
3.7 Opportunities for Social Entrepreneurs in India................................ 47
3.8 Economic Development and the Influence of Social Entrepreneurship 48
3.9 Conclusion 49
References.. 50

3.1 INTRODUCTION

Entrepreneurship is the process of identifying, assessing, and capitalizing on opportunities (Shane and Venkataraman, 2000). Entrepreneurship is the apparent driving force behind any nation's economic development, and it is the common denominator in all industrialized countries around the globe. The expansion of entrepreneurship is the single most crucial factor in a country's socio-economic development; as entrepreneurship increases, so do infrastructure and all other development indices. Richard Cantillon, a French economist, first used the term "entrepreneurship" in a business context in the 18th century, and he related it with risk and uncertainty. Additionally, entrepreneurship may be broken down into many distinct categories, one of which is Social Entrepreneurship (Gupta et al., 2020).

DOI: 10.1201/9781003256663-3

"Social entrepreneurship" has acquired considerable notice in the global arena in the last decade as a "new phenomenon" that is redefining how society thinks about creating social benefit (Makhlouf, 2011). This sort of entrepreneurship contains a wide range of diverse components, all of which are unique (Stirzaker et al., 2021). According to Johnson, 2000, social entrepreneurship "is developing as an ingenious way for addressing the complicated societal concerns." Rather than merely promoting change in society, social entrepreneurs serve as trailblazers for innovation in the social sector, demonstrating the entrepreneurial quality of a game-changing idea, their ability to build capacity, and their ability to concretely illustrate the quality of the concept while measuring social impacts (Surie, 2017). Individuals or teams (the inventive social entrepreneur) that have an entrepreneurial attitude and strong desire to succeed in creating new social value (Peredo and McLean, 2006) in the market and the society at large are the driving force behind social entrepreneurship (Perrini and Vurro, 2006). Societal problems account for the bulk of the work done by social entrepreneurs. As a first step, they organize the funds available to construct social structures in response to societal issues (Bulsara et al., 2013).

The expanding worldwide significance of the sustainability problem is another important reason in the expansion of social entrepreneurship. Due to global climate change and environmental impacts, achieving production and consumption systems that balance the interests of commercial, environmental and social stakeholders is becoming an essential policy goal in both developed and emerging nations (Porter and Linde, 1995). There have been rising business prospects for companies and governments to provide methods and services with a lower environmental impact, as well as stronger welfare conditions for workers, at the same time (Porter and Kramer, 2002). There has been an increase in customer concern about the negative effects of their choices on the environment and employees' welfare in developing nations. Firms are encouraged by these reasons to develop more sustainable production techniques to decrease reputation risks, target specific market niches, and cut costs.

Additionally, all three industrial revolutions, namely the first, second, and third, resulted in industrial prosperity, enhanced efficiency, and improved wellbeing in the nations that reaped the majority of their benefits. However, income equality within the industrialized nations that pioneered the industrial revolution was inequitable, even more so at the global stage, where inequality has emerged as a major issue alongside global warming and other sustainability concerns. Because the fourth industrial revolution, or the Fourth Industrial Revolution which is currently in the development stage, is predicted to bring a slew of benefits and challenges to countries' socioeconomic conditions, social entrepreneurs will be critical in balancing this shift (Morrar et al., 2017).

The primary goal of social entrepreneurship differs from that of traditional entrepreneurship in that it combines the pursuit of social good with the intent of financial gain. Most of the time, "social entrepreneurship" and "social service/work" are used interchangeably. Both have similarities, but one significant distinction is the existence of financial gain. It is not for profit but service that social workers and non-governmental organizations (NGOs) undertake their work in social work

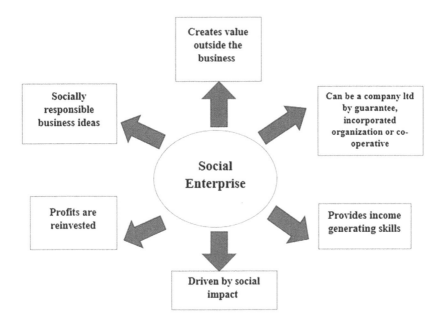

FIGURE 3.1   Characteristics of social enterprise.

Source: The author

(García-Jurado et al., 2021). On the other hand, social entrepreneurship focuses on non-personal advantages and incorporates revenues into social welfare. The bottom of the market pyramid is being addressed by social entrepreneurship, which offers products and services that are both innovative and cost-effective providing things to the underserved and making money from them. Figure 3.1 illustrates some of the major characteristics of a social enterprise.

3.2   WHAT DISTINGUISHES SOCIAL ENTREPRENEURSHIP FROM TRADITIONAL ENTREPRENEURSHIP?

In the philanthropic world, the term "social entrepreneurship" or "altruistic entrepreneurship" is widely used and often misinterpreted. Surdna Foundation's executive director Edward Skloot first used "Social enterprise" almost three decades ago. Altruistic entrepreneurship is the practice of spotting, assessing, and capitalizing on prospects that contribute to the well-being of society rather than personal or stockholder wealth. As a result, the most significant consideration in deciding on the organizational structure of a social enterprise is to determine how to mobilize the resources required to tackle a societal problem. As a result, the legal form does not define social entrepreneurship because it can be pursued through various techniques. There are social enterprises in the non-profit, commercial, and government segments (Austin et al., 2006).

When measuring success, social enterprise and traditional enterprise have very different measures. For example, traditional entrepreneurs use financial indicators like revenue and profit to gauge their success. In contrast, altruistic entrepreneurs use various measures to gauge their progress, including ecological, societal, and financial ones (Shukla, 2019). According to the type of social enterprise, some examples of these metrics include: A reduction in CO_2 emissions, land that has been sustainably managed, number of kilowatts of installed solar power, education is provided to low-income students, employment generation, etc.

It's possible to tell whether an enterprise's goals are primarily social or commercial by looking at the products and services it provides. From the mission, resource mobilization, funding, profit and wealth utilization, and financial performance, five significant distinctions can be made between Altruistic and traditional entrepreneurship.

- The primary goal of Altruistic entrepreneurship is to create public benefits for the benefit of society, while commercial entrepreneurship is primarily concerned with making money for the owner of the company (Austin et al., 2006).
- Mobilizing people and monetary avenues will be a defining difference between a social organization and a commercial enterprise, resulting in fundamentally different methods for managing financial and human resources.
- Many social entrepreneurs seek philanthropic funding during their first stages. Although these investors are looking for a return on investment (ROI), they are more likely to be interested in the company because of its social objective. On the other hand, a traditional company entrepreneur will typically seek funding from a venture capitalist firm—and they are solely concerned with the return on investment (ROI).
- An entrepreneur makes a profit and invests it in its growth and distribution of dividends. You join a business venture to make money and expand your wealth. Profits from social entrepreneurship ventures can be donated to charity or put to good use in various ways. Social entrepreneurship may not have any shareholders, and it is possible that the entrepreneur themselves will not gain much money from the venture. We must instead concentrate on our task.
- It is more difficult for a social entrepreneur to identify the effectiveness than a commercial entrepreneur since the social objective of the social entrepreneur is more difficult to quantify. An institution's connections with its many different financial and non-financial stakeholders, all of whom it must answer, become increasingly complex as the number and variety of those stakeholder relationships grows (Kanter and Summers, 1987). It is hard to evaluate societal transformation because of the lack of quantifiability, multi-causes, temporal elements, and perceptual differences in social influence. The difference between social and commercial entrepreneurship is listed in Table 3.1.

TABLE 3.1

Difference between Commercial and Social Entrepreneurship

Discussion Facts	Commercial Entrepreneurship	Social Entrepreneurship
Motivating Factors	Motivated by a desire to generate financial wealth	Inspired by a desire to create a more equitable and prosperous society/inclusive wealth (economic, social, environmental)
Beneficiaries	Entrepreneurs, shareholders, and investors all benefit from the wealth they have created.	The goal of wealth development is to benefit the community and society.
Metrics for determining success	Profits, turnover, market share, and other metrics are used to assess an enterprise's success.	The success of an organization is measured in terms of the positive impact it has on the lives of the people it serves.
Objectives	Concentrate on enhancing accumulated wealth by focusing on profit maximization	Focus on profitability to ensure the venture's autonomy and self-sufficiency.
Goals	Maximize their share of the market by overcoming competition from other businesses	Attempt to collaborate with other organizations to maximize the community benefit.

3.3 ATTRIBUTES OF SOCIAL ENTREPRENEURS

While the term "social entrepreneurship" has gained in popularity, the field is still in its infancy when compared to entrepreneurship as a whole. The success stories of people who have tackled complex societal problems are legitimizing the concept of social entrepreneurship. Famous organizations that are regularly featured in the context of Altruistic entrepreneurship include Ashoka (Mair and Marti, 2006), One Health, The Skoll Foundation and Schwab Foundation (Martin and Osberg, 2007). Societal concerns and the desire to improve well-being are at the heart of the appeal of social entrepreneurs (Zahra et al., 2008). The public frequently regards social entrepreneurs as heroes due to the plethora of social problems they address and the increased life quality they offer to afflicted populations. Social entrepreneurs are characterized by identifying a social issue and utilizing the community's resources to address it. It is usual for social entrepreneurs to have the following features.

Risk-taker: The establishment of a social entrepreneur entails risks and uncertainties, just like any other form of business.

- ***Dreamer:*** Dreaming about something that can be improved and mainly changed.
- ***Continuous Learning and Adaptability:*** Being a lifelong learner and receptive to new ideas are two essential qualities of a successful professional. Change is all they need if something isn't working for them. To stay on top of your industry, entrepreneurs must constantly adapt and modify to

stay ahead of the competition. When it comes to technology or customer service, they're always ready to adapt (Santhi and Kumar, 2011).

- **Visionary:** Social entrepreneurs identify the social issues and develop solutions that no one else can imagine.
- **Mission leader:** Those who work in social enterprises benefit from the guidance and leadership of social entrepreneurs.
- **Social value creator:** Social Entrepreneurs strive to improve the lives of others by taking into account all aspects of the impact they have on the economy, the ecology, and society as a whole. Figure 3.2 illustrates the essential characteristics of social entrepreneurs.

Some additional aspects of social entrepreneurs distinguish them from the rest of the entrepreneurs. True social entrepreneurs empower individuals and communities to rise above their current circumstances. Social enterprise is more than just charity work. Fig illustrates the motivation sources.

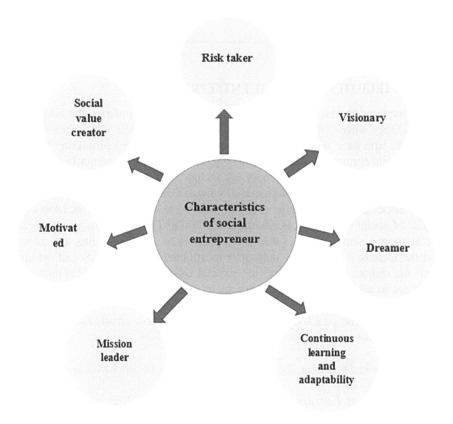

FIGURE 3.2   Characteristics of social entrepreneurs.

Source: The Author

3.4   THE SOCIAL ENTREPRENEURSHIP LANDSCAPE IN INDIA

India has a rich history of social entrepreneurs who have achieved great success. They have existed for ages, but the post-independent period has provided an ideal setting for their expansion. The failure of established institutions to address the problems of the poor and disenfranchised has spawned several grassroots initiatives and interventions that have been quite successful (Book). This created a level playing field for social enterprise. The rise of social entrepreneurship has impacted various industries in India, including agriculture, education, energy, finance, health care, housing, and sanitation. A growing number of social entrepreneurs have emerged as a result, with the potential to build long-term business models (ADB, 2012). An article authored by Bornstein and Davis (2010) cited various reasons for the significant prevalence of social enterprises in India. These include (a) the fact that India has a long history of non-profit organizations, (b) the fact that the nation has a vast population living below the poverty line and malnourishment, and (c) the fact that people in India have deep links to their families and communities, which provide a variety of resources to engage in social endeavors.

Vinobha Bhave was India's first prominent post-independence social entrepreneur, and Kurian of Amul Group was the second (Sinha, 2014). Both made significant contributions to the societies in which they established their operations. The earliest acknowledged recognition for social entrepreneurship in India came from Ashoka, which chose its first Indian fellow in 1982. Ashoka received valuable experience by participating in India, which aided it in demonstrating its worldwide footprint and articulating its capacity. There has been a steady flow of global philanthropic funds to India for the past few years. Wealthy donors with quick and long-term goals have recently increased their involvement in the local community. In the corporate sector, a new group of high-net-worth individuals is considering a charitable investment in the form of grants and Impact-focused investing. The field of strategic philanthropy, on the other hand, is still nascent in India (Swissnex India, 2015).

The practice of Corporate Social Responsibility (CSR) continues to be in the philanthropic arena in India but has evolved to include community development. CSR is growing more strategic as communities become more active and demanding due to global impacts. We must go beyond community involvement and the concept of philanthropy when considering a company's role in society by integrating corporate social responsibility into its primary business practices. The development industry now has a considerable opportunity to free up local wealth to invest in long-term social entrepreneurship and short-term social challenges. Mr. Deval Sanghvi, the President of Dasva, an entity that acts as the interface among those making investments in progressive transformation and those spearheading the changes, states that India's Altruistic entrepreneurship progress has significantly advanced over the past few years and that too many people are utilizing innovative business skills to build sustainable enterprises for both financial gains and non-profit purposes (Khanapuri and Khandelwal, 2011).

In countries that are still developing, social entrepreneurship is most beneficial. A developing country like India has its unique social problems, which serve as an appropriate starting point for entrepreneurial ventures. India's socio-economic inequities can be solved by social entrepreneurship. Several start-ups and new businesses have recently been founded to address social concerns in a long-term, sustainable way while also making money. Additionally, social entrepreneurship is essential in the Indian setting because most products and services are geared toward the wealthy, and those with fewer resources often go unmet because of a lack of resources. However, Social entrepreneurship depends on a supporting political, economic, social, and institutional climate to grow and thrive (Poon, 2011).

3.5   CHALLENGES FACED BY SOCIAL ENTREPRENEURS IN INDIA

The social entrepreneurship sector in India has its own set of challenges, just like any other business. GIZ (2012) conducted a study that uncovered some significant issues in the field. A few of these are: a dearth of a reliable means of finance for start-ups, inadequate access to industrial capital, elevated expenses for investors to locate and conduct due diligence on potential investment targets (particularly outside major cities), and low extensibility of businesses owing to shortage of funds, human capital issues, inadequate training, and an enabling environment (GIZ, 2012). Some of the most prominent challenges are elaborated in this section.

- In India, social entrepreneurship is often misconstrued with social work, and as a result, it is hard to establish itself as a distinct entity.
- The next issue that social entrepreneurs encounter is a shortage of originality when developing innovative ideas for improving society while also making money. This convergence is complicated to conceptualize and put into practice in India.
- Lack of financial resources is still a problem for Indian entrepreneurs. It is more difficult for social entrepreneurs to secure financial help from established financial institutions because they offer a unique product or service. A terrible scenario like this is a significant contributor to India's lack of social entrepreneurship (Rawal, 2018).
- Social enterprises usually don't offer jobs that pay well and offers bonuses which can be a challenge. Moreover, when it comes to social enterprise, the goal is to positively impact the community as a whole rather than making a profit for yourself. Thus, it's pretty challenging to get employees to work for the company in these conditions.
- Because social entrepreneurs care so much about societal change and the upliftment of people, they may choose to conduct their business less ethically. Unfortunately, this issue is rarely brought to light, yet it occurs in India in extreme circumstances (Rawal, 2018).
- Social entrepreneurs in India also face obstacles in their families. It's not easy to persuade someone to get into business instead of a job. There are many factors to consider when deciding if a company or one passed down through the generations is best. At this point, it's nearly impossible to persuade the families regarding the business idea (Santhi and Kumar, 2011).

3.6 PAST CONTRIBUTIONS OF SOCIAL ENTREPRENEURS IN INDIA

Some Indian instances of social entrepreneurship in various fields are listed below. In rural and semi-urban settings, most of this social entrepreneurship takes place. There is an increasing demand for social entrepreneurship as the needs of people are increasing.

- Harish Hande, is the pioneer of Selco, an entity that provides renewable energy sources to villages in our nation In India, this was the country's very first solar-funding project. A total of 25 functioning retail and service locations throughout the Karnataka region and more than 120,000 installations have been completed by Selco (Mollah, 2020).
- Urvashi Sahni the CEO, and founder of the Study Hall Education Foundation (SHEF), an entity that gives education to the poorest girls in India. Throughout her career, she has worked with over 900 schools, affecting the lives of over 150,000 girls directly and an additional 270,000 indirectly (Mollah, 2020).
- Ajanta Shah is the CEO and founder of Frontier Markets, a company that sells low-cost solar-powered products to rural India. The organization has now sold more than 10,000 solar systems, and they won't stop until they've illuminated every nook and cranny of the United States.
- An open-minded forum for social innovators, Ashoka Changemakers, was founded and is led by Sumita Ghose. Her purpose is to resurrect rural Indian creativity and craftsmanship so that he can give it the credit it deserves. It was a huge hit with the FabIndia retail chain, "Rangasutra," which she launched (Mollah, 2020).
- Goonj is a social entity that gathers unwanted cloths from the metro cities of the nation, sorts them, fixes them, and then gives them to the needy. Anshu Gupta created the organization. It is widely accepted that Goonj's relief efforts in Gujarat, Tamil Nadu, and Kerala during times of natural disasters were well appreciated (Bharech, 2021).
- Santosh Parulekar, the founder of Pipal Tree, a company that aspires to provide young people with formal training and legitimate employment opportunities around the country, founded the company. There are around 1,500 Pipal Tree employees who have been taught, and the company wants to open training centers across India shortly (Bharech, 2021).

3.7 OPPORTUNITIES FOR SOCIAL ENTREPRENEURS IN INDIA

Across the globe, social entrepreneurship is on the rise. Currently, there is a wide range of entrepreneurship prospects in child welfare, public safety, affordable healthcare, etc. Entrepreneurs in India can take advantage of today's knowledge-based economy. It is widely accepted that India has a vast pool of talented individuals who can start their businesses. Therefore, it is essential to put in the time and effort to create the appropriate conditions for aspiring entrepreneurs to succeed. These areas are critical for achieving this goal. Many pressing challenges in rural India can be solved through social entrepreneurship, an excellent opportunity for social entrepreneurship in rural India. However, there are several areas in which India has issues, and some of them are listed here:

- Solid and liquid waste management is unquestionably the foundation for a green and clean India. This is a priority. The present solutions are infrastructure-based, which means they demand significant investments and can never keep up with the rate of change. Social entrepreneurs can take advantage of this underserved market by developing new ideas and generating revenue (Jain, 2016).
- India's public administration has long struggled with the issue of child malnutrition. Despite decades of investment, India's child malnutrition rates are among the highest globally. Historically rooted causes such as poverty, injustice, and food scarcity have been blamed for India's child malnutrition epidemic. This is a topic that young social entrepreneurs may focus on and develop opportunities to solve.
- Water, energy, and healthy soil are essential to human existence on our planet. Compared to 30 years earlier, it was estimated that people were obtaining and utilizing more natural resources. In addition to causing deforestation, animal extinction, and depleting natural water resources, this level of consumption is also causing climate change. Renewable energy, particularly wind energy, can benefit from the efforts of social enterprises. Educating the public about renewable energy can also be done by social enterprises.
- Indian agriculture generates roughly 17% of the nation's Gross Domestic Product and employs more than 60% of the inhabitants, making it an important sector of the economy. Lack of affordable investment instruments, knowledge of high-quality inputs, ineffective innovation and financial data use, plus inadequate linkage with the market are all hindering the sector's development. To fill these gaps, social entrepreneurs might discover a wide range of options in agriculture.
- India is no longer an underdeveloped country, and unemployment is one of the characteristics of underdeveloped countries. Despite this, the government is plagued with a widespread problem of unemployment. Various causes contribute to the high rate of unemployment, which exacerbates concerns such as poverty, health problems, resource depletion, and antisocial behavior. The youth's enthusiasm is being diverted to nefarious actions because of unemployment. Youth and rural women's education and skill development can be improved through social businesses and social entrepreneurship.

(Agarwal et al., 2021)

3.8   ECONOMIC DEVELOPMENT AND THE INFLUENCE OF SOCIAL ENTREPRENEURSHIP

Social entrepreneurship, a fast-growing field that employs an innovative approach to produce an impact on society, is a necessity for economic growth and inclusion, particularly in poor nations. Social enterprises can create employment, provide new services and products, promote sustainability, and instill a sense of hope and optimism. On the other hand, social enterprises are not limited to developing countries (Kamaludin et al., 2021). For instance, in 2015, the social and solidarity economy sector accounted for 2 lakh firms, 2.34 million people, and 10.3% of total national employment in France, according to official figures. As a result, it contributed nearly 8% to the nation's gross domestic product (GDP) (Carree and Thurik, 2010). The social entrepreneurship industry provides three times as much to the GDP as farming, according to research released in 2018 by Social Enterprise UK (SocialEnt_UK, 2019). In comparison to what was originally believed, this is a major development. The research also revealed that the top five cooperatives in the United Kingdom—a social company—pay more tax than Amazon, Facebook, Apple, eBay, and Starbucks put together.

Social entrepreneurship frequently makes use of ethical activities including impact investing, conscious consumption, and CSR programs. When it comes to bringing about positive change in society through their projects, social entrepreneurs are typically eager to take on the risk and put up the effort necessary. Social entrepreneurs are frequently tasked with the responsibility of developing environmentally friendly products, providing services to underserved populations, and promoting humanitarian causes. One of the most prominent and noticeable outcomes of social entrepreneurship is the creation of new jobs, particularly for those who are less privileged, underserved, or marginalized in society. According to the Organization for Economic Cooperation and Development (OECD), social enterprises serve as "a link between unemployment and the open labor market."

3.9 CONCLUSION

Since the beginning of liberalization and globalization reforms in 1991, the Indian economy has proliferated. Due to rising social and environmental challenges, multidisciplinary approaches, and entrepreneurial energy must be used extensively in the social and environmental sectors. As a society's socio-economic conditions change, so do the opportunities for entrepreneurs. Since so many people in India are in need, the potential "market" for social initiatives is tremendous. Thus, Social Entrepreneurship has emerged as a new discipline that enables young people to sustainably create societal/economic value. In a country like India, where the organized sector employs only around 6–7% of the economically active workforce, there is an urgent need to redirect managerial expertise into projects that add/create value for the balance of the informal sector. However, the success in Social Entrepreneurship venturing is determined by many factors. It is exceptionally challenging to satisfy all the elements simultaneously as social entrepreneurship practitioners face time, money, manpower, internal capacities, and management efficiency constraints. Despite these hurdles, many notable instances of social entrepreneurship, such as Lijjat Pappad, Amul, and Gramin Bank, exist. In India, there is a great deal of opportunity for social entrepreneurs. The number of social entrepreneurs in India is rising, as are their efforts to develop low-cost remedies to society's myriad ills. Entrepreneurs in the social sector are being forced to evolve to keep up with the rapid speed of progress and competitive pressures. In light of the devastation caused by the COVID-19 outbreak, it seems likely that social businesses will require some form of government assistance in the coming months. However, there has been little mention of social enterprises in the announcements of financial packages to assist businesses, employees, and the self-employed. At this point, it is unclear whether or not this particular group will be granted special access to government funds during these unusual and risky circumstances.

REFERENCES

Agarwal, A., Gandhi, P. and Khare, P. (2021). Women empowerment through entrepreneurship: case study of a social entrepreneurial intervention in rural India. *International Journal of Organizational Analysis*, In press (ahead-of-print).

Asian Development Bank. (2012). India social enterprise landscape report. Retrieved 15 June 2022, from https://www.adb.org/publications/india-social-enterprise-landscape-report

Austin, J., Stevenson, H., and Wei-Skillern, J. (2006). Social and commercial entrepreneurship: Same, different, or both? *Entrepreneurship Theory and Practice*, 30(1), 1–22.

Bharech, A. (2021). 8 Amazing social entrepreneurs in India who are changing the face of urban India. https://digest.myhq.in/social-entrepreneurs-in-india/

Bornstein, D. and Davis, S. (2010). *Social entrepreneurship: What everyone needs to know.* Oxford University Press.

Bulsara, H. P., Gandhi, S. and Porey, P. D. (2013). Grassroots Innovations to Techno-Entrepreneurship through GIAN—Technology business incubator in India: A case study of nature technocrats. *International Journal of Innovation*, 1(1), 49–70.

Carree, M. A. and Thurik, A. R. (2010). The impact of entrepreneurship on economic growth. In *Handbook of entrepreneurship research.* Springer, 557–594.

García-Jurado, A., Pérez-Barea, J. J. and Nova, R. (2021). A new approach to social entrepreneurship: A systematic review and meta-analysis. *Sustainability*, 13(5), 2754.

Giz.de. (2012). Enablers for change – A market landscape of the Indian social enterprise ecosystem. Retrieved 15 June 2022, from https://www.giz.de/en/downloads/giz2012-enablers-for-change-india-en.pdf

Gupta, P., Chauhan, S., Paul, J. and Jaiswal, M. P. (2020). Social entrepreneurship research: A review and future research agenda. *Journal of Business Research*, 113, 209–229.

Jain, S. (2016). Opportunities for social entrepreneurs in India. *Entrepreneur.* www.entrepreneur.com/article/273849

Johnson, S. (2000). Literature review on social entrepreneurship. *Canadian Centre for Social Entrepreneurship*, 16(23), 96–106.

Kamaludin, M. F., Xavier, J. A. and Amin, M. (2021). Social entrepreneurship and sustainability: A conceptual framework. *Journal of Social Entrepreneurship*, 1–24.

Kanter, R. M. and Summers, D. (1987). Doing well while doing good: Dilemmas of performance measurement in non-profit organizations and the need for a multiple-constituency approach. In *The non-profit sector: A research handbook.* Yale University Press, 154–166.

Khanapuri, V. and Khandelwal, M. (2011). Scope for fair trade and social entrepreneurship in India. *Business Strategy Series*, 12(4), 209–215.

Mair, J. and Marti, I. (2006). Social entrepreneurship research: A source of explanation, prediction, and delight. *Journal of World Business*, 41(1), 36–44.

Makhlouf, H. H. (2011). Social entrepreneurship: Generating solutions to global challenges. *International Journal of Management & Information Systems (IJMIS)*, 15(1).

Martin, R. L. and Osberg, S. (2007). Social entrepreneurship: The case for definition. *Stanford Social Innovation Review*, 27–29.

Mollah, M. (2020). Top 10 most famous social entrepreneurship in India. https://mashummollah.com/social-entrepreneurship-in-india/

Morrar, R., Arman, H., and Mousa, S. 2017. The fourth industrial revolution (Industry 4.0): A social innovation perspective. *Technology Innovation Management Review*, 7(11), 12–20.

Peredo, A. M., and McLean, M. (2006). Social entrepreneurship: A critical review of the concept. *Journal of World Business*, 41(1), 56–65.

Perrini, F., and Vurro, C. (2006). Social entrepreneurship: Innovation and social change across theory and practice. In *Social entrepreneurship*. Palgrave Macmillan, 57–85.

Poon, D. (2011). The emergence and development of social enterprise sectors'. *Social Impact Research Experience Journal*, 33–56.

Porter, M. and Kramer, M. (2002). *The competitive advantage of corporate philanthropy*. Harvard Business, 1–14.

Porter, M. E. and Van der Linde, C. (1995). Toward a new conception of the environment-competitiveness relationship. *Journal of Economic Perspectives*, 9(4), 97–118.

Rawal, T. (2018). A study of social entrepreneurship in India. *International Research Journal of Engineering and Technology*, 5(1), 829–837.

Santhi, N. and Kumar, S. R. (2011). Entrepreneurship challenges and opportunities in India. *Bonfring International Journal of Industrial Engineering and Management Science*, 1, 14–16.

Shane, S. and Venkataraman, S. (2000). The promise of entrepreneurship as a field of research. *Journal of Management*, 25(1), 217–226.

Shukla, M. (2019). *Social entrepreneurship in India: Quarter idealism and a pound of pragmatism*. Sage Publications Pvt. Limited.

Sinha, D. K. (2014). Growth of entrepreneurship in India during Post-Independence. www.yourarticlelibrary.com/essay/growth-of-entrepreneurship-in-india-during-post-independence/40657

SocialEnt_UK. (2019). The hidden revolution. *The Hidden Revolution Social Enterprise UK*. www.socialenterprise.org.uk/policy-and-research-reports/the-hidden-revolution/

Stirzaker, R., Galloway, L., Muhonen, J. and Christopoulos, D. (2021). The drivers of social entrepreneurship: Agency, context, compassion and opportunism. *International Journal of Entrepreneurial Behavior & Research*, 27(6), 1381–1402.

Surie, G. (2017). Creating the innovation ecosystem for renewable energy via social entrepreneurship: Insights from India. *Technological Forecasting and Social Change*, 121, 184–195.

Swissnex India. (2015, June). Social entrepreneurship in India-unlocking unlimited opportunities. *Swissnex India*. https://www.swissnexindia.org/wp-content/uploads/sites/5/2016/05/SocialEntreprenuership-Report.pdf

Zahra, S. A., Rawhouser, H. N., Bhawe, N., Neubaum, D. O. and Hayton, J. C. (2008). Globalization of social entrepreneurship opportunities. *Strategic Entrepreneurship Journal*, 2(2), 117–131.

4 Creating Social Capital and Outcomes through Entrepreneurship in Industry 4.0 Perspective

Abhilash G. Nambudiri

CONTENTS

4.1 Introduction.. 53
    4.1.1 Types of Social Capital 55
4.2 Antecedents of Social Capital... 56
4.3 Moderating Role of Social Entrepreneurship Model.................................... 60
4.4 Methodology... 60
    4.4.1 Population 60
    4.4.2 Sampling Design... 61
    4.4.3 Tool for Data Collection 64
    4.4.4 Translation and Validation of Translation 64
    4.4.5 Data Collection 65
    4.4.6 Data Analysis and Findings 66
4.5 Results and Discussion 67
4.6 Conclusion 70
4.7 Limitations & Scope for Future Research 71
References... 71

4.1 INTRODUCTION

Social Entrepreneurship is an important cog in the wheel when we discuss Industry 4.0. The issues of trust and inequalities are the important social issues related to the discussion on social entrepreneurship and Industry 4.0. Through the application of entrepreneurship principles to social issues, we can tackle these issues. Social capital is regarded as a proxy for the amount of trust in any society. Well-being of people and the country depends upon the trust prevailing in the society, and it has a role to support social and economic development. The trust between individuals and trust in institutions acts as the determinants of economic growth, and cohesive well-being. Trust is one of the major dimensions of social capital, and research shows that this dimension cannot be neglected for social and economic development (Algan and

DOI: 10.1201/9781003256663-4

Cahuc, 2010). This chapter focuses on the role of "social capital" for progress of society and well-being.

Social enterprises can deploy the community asset of social capital to effectively overcome the operational implementation challenges (Rydin and Holman, 2004). The view of the Social Capital Theory is that, social relationships among people in a community can be a productive resource. In the early studies on social capital, it is pointed out that the social capital enables coordination and cooperation for mutual benefit within a community. Social capital is defined as "The sum of the resources actual or virtual, that accrues to an individual or a group by virtue of possessing a durable network of more or less institutionalised relationships of mutual acquaintance and recognition" (Bourdieu, 1986). Because of the productive nature of these social relationships, they would often act as a catalyst in the process of creating benefits to the society (social benefits).

L. J. Hanifan was the first one to use the term "social capital" (Hanifan, 1916). In this study, social capital is referred to the after effect of efforts in making people of a community to come and act together towards a common goal. Social capital which is defined as an informal access to others' resources through social ties is not a monolithic concept, it comprises of three distinct types viz., bonding capital, bridging capital and linking capital (Darcy et al., 2014; Chen et al., 2015). These different types of social capital could perform different functions in creating of social outcomes by social enterprises. Bourdieu conceptualized social capital as a variable at the individual level suggesting that an individual could acquire social capital through purposeful actions and, like financial capital, social capital could be transformed into conventional economic gains. His emphasis was on the connection of networks and sustained relationships that emerged from those networks. The concept of social capital was later made more popular through a book "Bowling Alone" by Robert Putnam and kindled interest for further research on this topic (Putnam, 1995).

Coleman, another early social capital researcher, also linked social capital to enhanced performance. He defined social capital is defined as a set of socio-structural ties which an individual could mobilize (Coleman, 1990). Key to the concept which Coleman proposed was that, some outcomes could only be realized with the existence of an embedded social capital among individuals. Coleman associated social capital with both individual and collective assets based upon structural relationships with others.

From the previous studies it was theorized that the concept of social capital enhances the outcomes of actions. This theory was based on the fact that social capital is an investment in the social relations which is expected to give tangible returns in the marketplace. Social capital focuses on the embedded resources of one's social network and how the person can access and such resources for personal benefits.

Social capital leads to the economic development of a country level (Doh and McNeely, 2012). Investments in social infrastructures would help any particular community to address common environmental challenges (Bernard et al., 2014). This calls for governments to concentrate on projects, which is likely to correct inefficiencies in resource distribution in a community and drive the process of accumulation of social capital through appropriate policy formulation. It was observed that

a. For governments to have a significant part in creating social capital, they need to formulate appropriate policies that would be intended to correct inefficiencies of resource distribution, which in-turn would accumulate social capital.

b. Utilization and effectiveness of national budget allocation has improved in those places where the government had supported projects on land improvement. It was also found that, while the projects contributed significantly to national land conservation process, social mobility also had improved with the nation-wide accumulation of social capital (Yamaoka et al., 2008).

Social capital is a true contextual construct, which can be best studied through perception-based approaches (Subramanian et al., 2003). Previous studies in this filed had measured social capital using three dimensions viz. structural dimension, cognitive dimension and relational dimension (Li et al., 2014). Structural dimension talks about the bondages existing within a community, which any embedded actor of the community has access to. It is expressed through the existence of network ties among social entities (Yim and Leem, 2013). Increased interaction leads to increase in the structural dimension. Relational dimension talks about the personal relationships created by repeated interactions (Krause et al., 2007). Trust is a key factor in explaining relational capital. Cognitive dimension is referred as the existence of shared meanings among embedded actors of a community (Nahapiet and Ghoshal, 1998). Apart from these three dimensions, past studies had identified four more dimensions: togetherness, group characteristics, neighbourhood connections, and volunteerism even though with overlapping meanings with the above said three dimensions (Narayan and Cassidy, 2001). Since some of these dimensions have overlapping meanings with other dimensions, it was suggested not to use a dimension-based approach while exploring the antecedents of social capital. Hence, this study tries to measure social capital as the above referred three distinct types in order to attain its objective.

4.1.1 TYPES OF SOCIAL CAPITAL

There are generally the three types of social capitals explored below:

a. ***Bonding capital:*** Bonding social capital is defined as the capital derived out of relations within homogenous groups like family or friends. It is often equated with "strong ties." These ties are important for "getting by" (Chazdon et al., 2013). Participation in geographic-based networks create bonding capital and helps in accelerating activities of political action groups (Hays, 2015). This in-group connectedness helped members to cope with challenges in socio-economic environment such as access to loan for poor or people with inexistent collateral assets (Woolcock and Narayan, 2000).

Social cohesion and homogeneity positively affect the life satisfaction in a community (Cheung and Leung, 2011). However, when people tend to access only to this type of social capital, there are less probability that they escape from their existing unfavourable social condition. Subsequently, the network for an individual becomes a closed one as bonding capital would become exclusive. This results in reduced access to other resources possessed by other groups, which subsequently results in lowering of social capital endowment.

b. **Bridging capital:** Bridging capital is defined as the capital derived out of networks with relatives, associates or colleagues. These ties are weaker but are important in "getting ahead" (Chazdon et al., 2013). Bridging capital, also referred as "weak ties" help people to get access and garner support from other groups, hence it is helpful in mobilizing other forms of capital effectively (Nichols and Fernandez, 2007; Nicholson et al., 2016). The exchange of knowledge and information between such groups allow the communities to benefit from different ways of social endowment accumulation and, result in more social capital formation.

c. **Linking capital:** It is defined as the capital derived out of ties between individuals and groups belonging to different social, economic or power strata (Chazdon et al., 2013). Linking capital denotes the connecting ties between individuals or groups in positions of different political, social or financial superiority (Sabatini, 2009).

Looking at all the literature evidence regarding social outcome and types of social capital, this research study postulates that all the three types of social capital have a significant relationship with the dependent variable, social outcome.

4.2   ANTECEDENTS OF SOCIAL CAPITAL

Past researchers have identified many variables that can serve as the antecedents to social capital (Richey, 2007). Ethnicity or sense of belongingness to a particular group was found to be predictor of social capital (Darcy et al., 2014; Kitchin and Howe, 2013; Shoji et al., 2014; Barnes-Mauthe et al. 2015). Homogeneity was also found to be a significant predictor for social capital (Cheung and Leung, 2011; Coffé, 2009). Nisha et al., in her research has stated that information-sharing can impact creation of social capital (Paul et al., 2016). Researchers have also suggested the significant influence of community participation on social capital (Campos et al., 2015; Stuart, 2013; Hays, 2015; Katherine et al., 2010; Neal and Neal, 2019; Rasoolimanesh et al., 2017; Robin et al., 2005). The civic engagement was cited as a significant determinant in creation of social capital (Svendsen and Srensen, 2006). Social Identity was also found to be a significant predictor as per several past researches in this area (Li, 2007; Pinho, 2013). Lawson suggested that increased levels of communication can also create social capital in a business environment (Lawson et al., 2008). Past researches also suggested the significant influence of altruism or pro-social motivation on creation of social capital. A comprehensive list of the antecedents established through previous studies is given in Table 4.1.

TABLE 4.1
List of Antecedents of Social Capital

Sl No	Antecedent Variable	Context	Description
1	Sense of Belongingness	Social organization in sports (Darcy et al., 2014)	Organisational bonding—>[1] sense of belongingness—> bridging capital
		Rural sports organizations (Tonts, 2005; Kitckin et al., 2013)	Trust reciprocity—> sense of identity (belongingness) —> bonding capital
		Low-income school community (Shoji et al., 2014)	Sense of belongingness—> social capital
		Resident communities (Richie, 2007)	Program to create community belongingness—> trust—> social capital
		Social network analysis on fishermen community (Barnes-Mauthe et al., 2015)	Ethnicity (sense of belonginess to a community) —> social capital
2	Homogeneity	Municipal bodies (Coffe, 2009)	Lower homogeneity—> low social capital
		Resident communities (Cheung et al., 2011)	Homogeneity—> social capital
3	Information Sharing	Supplier innovation on firm's innovation capability (Kulangara et al., 2016)	Information sharing—> social capital
4	Community Participation	Grass-root organizations (Hays et al., 2015)	Community participation—> social capital
		Measuring social capital in NGOs (Campos et al., 2015)	Community participation—> social capital
		Organized youth programs (Robin et al., 2005)	Activities which promoted participation—> social capital
		Community health program (Stuart, 2013)	Participation—> bonding and bridging capital
		Urban resident communities (Neal and Neal, 2019)	Intra community interaction and participation—> organizational social capital
		Neighbourhood communities (Katherine et al., 2010)	Community participation—> social capital
		Social organizations (Rasoolimanesh et al., 2017)	Community participation—> social capital
5	Civic Engagement	Rural Danish civic society (Svendsen et al., 2006)	Civic network engagements—> bonding capital
6	Civic Norms	Entrepreneurship (Doh and McNeely, 2012)	Civic Norms—> trust—> bonding capital
7	Social Identity	Social capital scale development (Pinho, 2013)	Social Identity—> bonding and bridging capital
		Migrant communities (Zetter et al., 2006)	Identity—> social capital

(Continued)

TABLE 4.1 *(Continued)*

Sl No	Antecedent Variable	Context	Description
		Theory building research (Li, 2007)	Strong tie (identity) —> Strong social capital
8	Pro-Social Motivation	Entrepreneurship (Kibler et al., 2019)	Negative relationship between prosocial motivation and linking capital
		Health care (Henny and Marco, 2013)	Pro-Social Motivation—> trust—> social capital
		Parenting (Aydinli et al., 2015)	Prosocial Motivation—> volunteering—> social capital

[1]—> means "leads to"

Looking at the relevance of the antecedent variable in the context of this study, its frequency in the past literature and the novelty of variables (newness), this study is considering Sense of Belongingness (sob), Social Identity (si), Community participation (cp), and Pro-Social Motivation (psm) as the antecedent variables to understand the construct of social capital.

a. *Sense of Belongingness:* Sense of belongingness is defined as the self-conception defining the features of self-inclusive social category. Social researchers had proved that sense of belongingness, which symbolizes the identification of people to a community, has a significant positive impact on the quantity of shared knowledge, thereby creating a better shared vision (Chiu et al., 2006). Through a study conducted among start-up firms concluded that the sense of affiliation to a particular business or industry cluster influences the formation of social capital (Presutti and Boari, 2008). In study in the context of micro-credit and agriculture production and marketing, it was found that sense of belonging positively affects the formation of social capital and they have suggested to explore this relationship further for better understanding (Seferiadis et al., 2015). Based on findings from the existing knowledge body, this study is postulating that sense of belongingness will have a significant impact on all types of social capital.

b. *Social Identity:* Social identity is defined as the main feature of the individual's identification with the community. Potential sources of social capital based on identity are not limited to actual ties and include potential future social ties. However, individuals in the same identity groups are more likely to develop future relationships with each other than with non-members because members of the same identity group are similar and easily attracted to each other. A study on community social capital pointed out that it is derived from actual and potential relationships based on similar shared cognitive representations (Oh, 2000). Further the immigrants from the same country, although strangers to each other, will be more likely to trust or help each other than trust or help strangers from other countries. Based on findings from the existing knowledge body, this study is postulating that social identity will have a significant impact on all types of social capital.

c. ***Community Participation:*** Community participation is defined as the instance of a person being a participant in any community activities (Jaafar et al., 2017). It is one of the important aspects which create social capital. Adams et al. in their research to develop a tool to measure bridging social capital has identified community participation as a factor affecting creation of bridging capital (Villalonga-Olives et al., 2016). It was also found that through responsive and reciprocal communication, creating shared experiences and institutional linkages, social interactions get enhanced in a community, in turn leading to the social capital creation (Shoji et al., 2014). A research study on social capital involving Brazilian adolescents noted that community participation was a significant discriminating factor in segregating respondents with different levels of social capital. The research indicated that there is a positive correlation between amount of community activities and social capital created (Campos et al., 2015). Through an action research on the conservation of world heritage site and tourism development, it was observed that the spontaneous participation by the beneficiaries led to creation of trust and ownership which is an indication of existence of social capital in the community (Jaafar et al., 2017). Successful implementation of social projects and developmental initiatives draw its major success factors from the enhanced social interaction (Park et al., 2017). One of the main reasons for this enhanced social interaction is the community participation (Park et al., 2017; Rasoolimanesh et al., 2017). Based on findings from the existing knowledge body, this study is postulating that community participation will have a significant impact on all types of social capital.

d. ***Pro-social Motivation:*** Pro-social motivation is defined as the desire to expend effort in order to help other people (Grant and Sumanth, 2009). Pro-social motivational helps to improve voluntarism which is an indicator of social capital, in an organizational setting (Aydinli et al., 2015; Grant and Sumanth, 2009). Putnam had suggested that altruism (volunteerism) is an indicator to social capital (Putnam, 2000). Harper, in a study to compute social capital index observed that, measure of community volunteerism was highly correlated with social capital index (Harper, 2001). In a research involving 570 parents and non-parents from Turkey, it was found that, prosocial motivation resulted in parents' volunteering propensity (Aydinli et al., 2015). Pro-social motivation would also positively affect organizational and social outcomes. In a meta-analysis involving managerial behaviour and organizational social capital, it was found that in organizations it was important for generating social capital (Pastoriza et al., 2007). Consolidating the findings from past researches, this study proposes the following hypotheses which need to be empirically tested. Based on findings from the existing knowledge body, this study is postulating that pro-social motivation will have a significant impact on all types of social capital.

4.3   MODERATING ROLE OF SOCIAL ENTREPRENEURSHIP MODEL

There were earlier studies which tried to segregate social enterprises into different categories based on the business models. Santos et al. classified social projects into four as market hybrids, bridging hybrids, blending hybrids, and coupling hybrids. Market hybrids are closer to pure commercial businesses. Such enterprises are different from a commercial venture only in its mission, which is predominantly social in nature. Blending hybrids serve clients with paying capacity who also are their beneficiaries. Bridging hybrids attend to clients and beneficiaries as two different entities. Bridging hybrids are forced to bridge the needs and find required resources to do so for both entities. Coupling hybrids also similar to bridging hybrids, but they differ in the creation of spill-over effects. While for bridging hybrids spill-over effect happen automatically, for coupling hybrids distinct social interventions is required alongside its commercial operations. It was also observed that profit alignment with social impact can drive long-term competitive advantage of social businesses (Santos et al., 2015).

All the above arguments suggested that business models can be used as a meaningful parameter to classify social entrepreneurships. There are social enterprises which predominantly thrive on external funding sources for their existence and financial sustainability. These social enterprises can be classified as funded model of social entrepreneurship. However, there are social enterprises who support its existence and finances its activities through revenue generated by itself. They operate within the framework of "free-market" either through the sale of products or services they had created. They can be classified as market models of social entrepreneurship. Market models and funded models are the two extremes of a business-model continuum, and social enterprises based on other possibilities of mixed models are also possible. This study explains only the two extremes of business models viz market model and funded model. The conceptual model of social capital and social outcome creation is depicted in Figure 4.1.

4.4   METHODOLOGY

The study wanted to focus on the process of creation of social outcome rather than understanding "why" it is happening. Hence, descriptive methodology was found to be the most suitable for this study. The study adopted a descriptive methodology to understand the process of social outcome creation through the creation of three different types of social capital.

4.4.1   POPULATION

Since the social capital creation was hypothesised to be moderated by the model of social entrepreneurship, this study considers individual social enterprise as the unit of analysis. As explained earlier, this study classifies social entrepreneurship into two categories: funded model and market model of social entrepreneurship. For consistency of samples chosen for this study, there an inclusion criterion for any social enterprise to qualify as a sample for this study. This study only included enterprises that

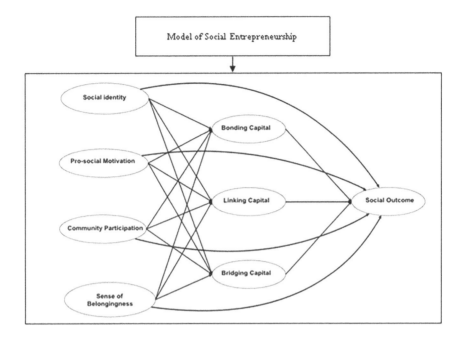

FIGURE 4.1   Conceptual Model of Social Capital and Social Outcome Creation.

completed all necessary regulatory compliances for its inception and were in existence for at least two years. Any recently launched social enterprise, which was less than two-year-old were not considered fit to be included in the sample for this study. Thus, the population of enterprises under consideration for data collection was described as "any social enterprise formed in the state of Kerala (southernmost state in India) and is active for a time period of at least two years as on date of data collection."

There were around 25,000 such "enterprises" which would qualify the criteria of inclusion for this study at the time of data collection. In the state of Kerala, some of these entities collectively could be put under one single social entrepreneurial initiative, but for all practical reasons, the context and locational distinctiveness would make each of the "projects" unique. Hence, this study considered each project as a separate data collection unit. A list of projects considered as the population for this study is given in Table 4.2.

4.4.2   SAMPLING DESIGN

The study used a convenient sampling design. Questionnaires to collect data were circulated to social organizations and responses were collected from both beneficiaries and implementing agency personnel on variables chosen for the study.

  a. **Sampling Frame:** The study needs to take samples from the population of both funded and market types of social projects implemented in Kerala. A

TABLE 4.2
List of Projects Considered as Population

Sl. No	Project Details	Numbers
	Major Funded Projects	
1	Jalanidhi—World bank aided rural water supply project	5883*
2	Orphanages (Both state run and privately run by individuals or charitable trusts)	442**
3	Old age homes (Both state run and privately run by individuals or charitable trusts)	176**
4	Housing Schemes of line departments of Government of Kerala	312**
6	Community water supply schemes implemented through line departments and Local Self-Governing bodies of Government of Kerala (LSGD).	143**
7	Tribal Welfare schemes of Government of Kerala	102**
8	Social Afforestation programs by Dept of Forest, Government of Kerala	67**
9	Western Ghat Development Scheme (implemented through LSGD)	157**
10	National Watershed Development Program for Rain-fed Agriculture (NWDPRA)	148**
	Major Market Projects	
11	Akshaya ICT kiosks	2600***
12	Kudumbashree projects (Micro-Enterprises)	15000****
13	Farmer Producer organizations under NABRD	159**
14	Co-operative agri/food processing centres6	76**
15	Agri Cooperatives	102**

* Exact number given as per official website (https://jalanidhi.kerala.gov.in)

** Approximate number arrived after scanning though various department and scheme related reports and discussion with officials

*** Approximate number of ICT kiosks as per official website (http://akshaya.kerala.gov.in/about)

**** Approximate figure of non-farm micro enterprises as per official website (www.kudumbashree.org/pages/653)

consolidated list for projects considered as population is given in Table 4.2. Since it was difficult to obtain an exhaustive list of individual projects under each category (sampling frame) given in Table 4.2, the study resorted to a non-probability sampling method. An approximate quota for each type of project was assigned based on their proportion in the population. Based on the criteria of impact, reach, and availability of data, the study created a sampling plan as described below.

b. *Funded projects:* Based on the contextual relevance of antecedent variables, scope of project impact on communities, and prospect of easy access to data and information, the study chose to collect data from four funded type of social projects viz. *Jalanidhi* projects, old-age homes, orphanages and privately funded charity projects. *Jalanidhi* projects, old-age homes, orphanages were chosen because, they were the three among the top four contributors to the list given in Table 4.2. The study avoided housing schemes of government of Kerala implemented through Kerala State

Housing Board because, it was very difficult to get access to government data regarding housing. The study also chose to consider private charity activities of individuals as part of sample as it fitted well within the inclusion criteria of this study. However, as there was no comprehensive list of these projects, the study restored to a non-probability-based quota sampling. All the chosen projects were similar in the aspects of funding pattern, which required external funds for implementation and sustainability of projects. The level of funding required might vary within this selection criteria, but all of the selected projects were predominantly banking on external funding sources for successful running and hence could be clubbed under the purview of funded projects.

Jalanidhi projects happened in three phases starting from 2003 in Kerala. *Jalanidhi* has covered 227 grama panchayats with over 7,750 beneficiary groups, 5,883 schemes and covered over 2.5 million beneficiaries of Kerala. There were 61 supporting social organizations, which undertook the implementation support of these 5,883 schemes. From the available pool of 1,168 water supply schemes completed by these 12 supporting organizations, a sample of 110 water supply schemes were selected for data collection. There are 14 government run and 162 private charitable trust run old-age homes in Kerala. There are 29 government run children homes and 413 private charity run orphanages in Kerala. The study had access to information regarding 15 such projects where there was an instance of private charity involved and the size of beneficiaries involved where not less than 25. Most of it were distribution of school kits and festival kits to the needy people. Beneficiaries of projects provided responses for the constructs of three types social capitals and also on all antecedents of social capital viz., sense of belongingness, social identity, pro-social motivation, and community participation. Conclusively all individual responses corresponding to one project were corroborated into one data point for the purpose of data analysis.

c. *Market Projects:* Based on the contextual relevance of antecedent variables, scope of project impact on communities and prospect of easy access to data and information, the study decided to collect data from three market type of social projects viz. *Akshaya* e-centres, *Kudumbashree* non-farm enterprises, and Farmer Produce Organisations (FPOs). These three projects were chosen because cumulatively they accounted for more than 95% of the market model projects identified in the population. All the chosen projects were similar in the aspects of market dynamics governing its sustainability, i.e., all chosen projects generated revenue internally without the help of any external funding sources for the long-term sustainability. Akshaya makes information accessible, provide transparency to the governance process, and fuel overall socio-economic growth. According to the *Kudumbashree* sources, there were around 15,000 non-farm based micro enterprises spread across the state of Kerala. All of them were involved in income generating activities through Self Help Groups (SHGs) and were self-sustaining by itself. Kerala have a total of 159 FPOs with over 53,000 shareholders spread across 14 districts. They operated in accordance with the market dynamics and were self-sustaining business models to generate surplus.

d. *Unit of Observation:* Since the "project" was considered the sampling unit for this study, and getting all beneficiaries of a single project to respond to the data collection process was not practical. Hence, the researcher decided to select two beneficiaries to respond to the data collection process per project. From each of these two beneficiaries of the same project, data was collected regarding three types social capitals viz. bonding, bridging, and linking capital and also on all antecedent variables of social capital viz., sense of belongingness, social identity, pro-social motivation, and community participation. These two responses were later clubbed into one data point during the data analysis phase by calculating the mean value of two responses.

e. *Sampling Method and Sample Size:* From the sample frame as explained above, the study chose projects proportionately using the quota-sampling technique. The sample-size needed for the study was estimated based on the sampling design. The focus of the study was to understand the process of creation of social capital and its variation across two different types of social entrepreneurship models. The variance was assessed using multi-group-invariance testing procedure in Structural Equation Modelling (SEM) (Byrne, 2010). For this study, sample size estimation considered the stringent norms of SEM. This study followed Bonett's and Bentler's recommendations of at least 400 samples or a ratio of 1:5 subjects per variable (Bentler and Bonett, 1980). The researcher targeted to collect more than 400 samples with a ratio of 1:9 subjects per variable.

4.4.3 TOOL FOR DATA COLLECTION

A schedule was used to collect data corresponding to each project. The questionnaire was designed by incorporating individual scales of measurement of each variable as prescribed by the respective studies. There were 52 items in the original questionnaire, which was reduced to 46 after initial validation of scales using a pilot study.

All scales used to collect data in this study were adapted from published articles of past studies after looking into the adaptability to the context of current research objectives. A snapshot of variables and the literature from which it is adapted is given in Table 4.3.

4.4.4 TRANSLATION AND VALIDATION OF TRANSLATION

Since the context of data collection was regional, it was necessary to translate the tool to Malayalam (local vernacular language). It was needed to confirm that translated questionnaire adequately captures description of tool's original items which was in English. The study checked that the translated version of the questionnaire was readily understood by subjects as envisaged by its original English version. It was also confirmed that, while translating and administering the tool, the validity and reliability of the tool was not affected. Several methods could be used to validate translated questionnaire; none of which is fail-safe (Sperber, 2004).

TABLE 4.3

Variables and their Sources

Sl No	Variable	Scale Adapted from
1	Social Outcome (so)	Urban, 2015
2	Bonding Capital (boc)	Chazdon et al., 2013
3	Bridging Capital (brc)	Chazdon et al., 2013
4	Linking Capital (lic)	Chazdon et al., 2013
5	Sense of Belongingness (sob)	Chiu et al., 2006
6	Social Identity (si)	Pinho, 2013
7	Pro-Social motivation (psm)	Grant and Sumanth, 2009
8	Community Participation (cp)	Rasoolimanesh et al., 2017

4.4.5 DATA COLLECTION

Questionnaire with items measured on five-point Likert scale was used for data collection process while administering the data collection process. Mailers (both electronic and physical) were sent to all supporting organizations of *Jalanidhi*. Only 12 of them responded to the study's request to facilitate data collection. Samples for other funded projects were chosen as per the judgement for data collection. Samples for *Akshaya* project, *Kudumbashree* projects were also selected proportionately as per judgement. All FPOs were contacts through e-mail and post, and four of them responded positively to co-operate with data collection.

Data collection was conducted using structured interview with the help of the pretested and validated questionnaire. All the respondents were briefed about the objectives of the study in their native language and consensus was taken from each of them before extracting any response. Before the discussion, each respondent was asked to relate all their responses to the identified social project, which they were a part of. By this, subjective understanding-bias from the side of respondents was minimized.

Two beneficiaries from each project gave their responses based on their interactions with the social project. These two data points were later collated into one data point. Beneficiaries provided responses for the constructs of three types social capitals viz. bonding, bridging, and linking capital and also on all antecedents of social capital viz., sense of belongingness, social identity, pro-social motivation, and community participation. Since social outcome of a project was operationalized by its four aspects viz., reach, innovativeness, scalability, and sustainability, this study found it apt to collect responses for the social outcome variable from the office bearers or decision-makers of the social enterprises who implemented the projects.

The study collected data from a total of 195 funded social projects spread across Kerala. Data from 110 *Jalanidhi* projects, 45 orphanages, 25 old-age homes and 15 privately funded charity projects were collected on all the seven constructs relevant to the study. Data was also collected from 218 market model social projects. This included 65 *Akshaya* centres, 149 non-farm *Kudumbashree* micro enterprises and four Farmer Producer Organizations (FPOs). In total the study collected data from 413 projects combining both market type and funded type of social project together. A snap-shot of sample details is given in Table 4.4.

TABLE 4.4
Snapshot of Samples

		Project							Total
		Akshaya	KudumbaShree	FPO	Jalanidhi	Orphanage	Old-age home	Other	
Intervention	Funded	-	-	-	110	45	25	15	195
Model	Market	65	149	4	-	-	-	-	218
Total		65	149	4	110	45	25	15	413

4.4.6 DATA ANALYSIS AND FINDINGS

The study received 413 usable data points collected as detailed in the methodology section. This section would explain various tests that were carried out in-order to answer the hypotheses outlined in this study.

a. *Checking consistency and reliability of the tool:*Internal consistency was checked by computing Cronbach's Alpha for each variable under consideration (George and Mallery, 2010). All values of Cronbach's Alpha were above the threshold limit of 0.7 and ranged within 0.7 to 0.81. Reliability of the tool was established using Composite Reliability (CR) statistics (Werts et al., 1978). All CR values were also found to be within the permissible limit i.e., >0.7. CR values ranged from 0.81 to 0.92. Table 4.6 shows the details regarding Cronbach's Alpha and CR of each variable under consideration.

b. *Analysis:*Analysis was carried out in two stages. First, the combined data set data was subjected to model testing without bifurcating into two models. This was done to assess the model-fit and direct relational hypotheses as per the proposed conceptual model. Subsequently, the data set was split into two; market model and funded model data-sets and analysed separately to test mediation and moderation hypotheses. The covariance-based structural equation modelling (SEM) technique was used for data analysis. The measurement model was constructed using all the variables of the study. Confirmatory Factor Analysis (CFA) was carried out to arrive at fit indices of the measurement model. The fit indices values of the measurement model for the combined data set is given in Table 4.5. The table indicates that RMSEA (< 0.08), GFI (> 0.9) and SRMR (< 0.08) values were well within the acceptable range and hence provided good fit between the proposed model and the dataset.

The study went on to check the moderation effect of social entrepreneurship models on the proposed model for creation of social outcome and found that the model behaves significantly different under two different models of social entrepreneurship viz. market and funded models. Generally accepted method of Multi Group Invariance testing procedure was carried out to arrive at the above said conclusion (Anderson and Gerbing, 1988). Subsequently the model was tested using two

TABLE 4.5
Measurement Model Fit Indices

$\chi 2$	Normed $\chi 2$	SRMR	GFI	NFI	TLI	CFI	RMSEA
2230.88	1.46	0.077	0.893	0.862	0.923	0.949	0.033

TABLE 4.6
Cronbach's Alpha and Composite Reliability (CR)

Variable	Items/ Dimensions	Cronbach's Alpha (Internal Consistency)	CR (Composite Reliability)
Social Outcome	11/4	0.70	0.92
Bonding Capital	7/2	0.78	0.84
Bridging Capital	7/2	0.70	0.89
Linking Capital	11/2	0.81	0.91
Sense of Belongingness	4/1	0.70	0.81
Social Identity	8/2	0.73	0.89
Pro-social Motivation	5/1	0.74	0.84
Community Participation	4/1	0.70	0.82

different data sets on funded and market model separately and the path coefficients resulted are tabulated in Tables 4.7 and 4.8.

4.5   RESULTS AND DISCUSSION

The study found that only bonding capital was influencing the social outcomes significantly. It was also found that the process of social outcome creation was routed more through the creation of bonding type of social capital with three out of four antecedents influencing bonding capital and bonding capital having high significant influence on creation of social outcome. It was also observed that, in funded model of social projects, more antecedent variables were having direct, significant and positive relationship on the process of creating social capital than in a market model. Both community participation and sense of belongingness were found to have direct and significant relationship with social outcome.

The study also looked at the way social capital was being formed in two distinct types of social projects. In funded type of social project; it was observed that, prosocial motivation, social identity and sense of belongingness were having significant positive relationships with bonding capital (p-values 0.000, 0.004, and 0.017 respectively). Social identity was having significant influence on creation of bridging capital (p-value of 0.016). Community participation had significant positive impact on creation of linking capital (p-values of 0.000). But pro-social motivation was found to have a significant negative influence on linking capital (path coefficient of -0.235). The study identifies that, the more an entity is associated with altruism or pro-social behavior, he/she would be more glued to the project and would concentrate less on creating ties outside the project scope, especially when the project is of a funded type tackling issues related to civic engagement. This would result in reduced scope for the creation of bridging and linking kind of social capitals. For a market model social project, the antecedent pro-social motivation was found to have significant positive impact on bonding capital and bridging capital (both with p-value of 0.000). Community participation and social identity were found to have significant impact on linking social capital (both with p-values 0.000). Pro-social motivation was found to have negative significant impact on linking capital with a path coefficient of -0.235.

Based on the results of path analysis of moderated models, as given in Tables 4.7 and 4.8, this study arrived at the following conclusions. There is a significant moderation effect of the project type on the proposed model for understanding the process of social outcome creation. For a market model of social project, it was found that the major focus was on the creation of outcomes and the social capital formation became only secondary. In market model of social intervention project, two out of three types of social capital that existed in the society or formed as a result of the project were influential in creating social outcomes. The path coefficients of bonding capital, and

---

TABLE 4.7

Standardized Regression Path Coefficients and Effect of Moderation: Between-Social Entrepreneurship Models

Path	Funded Model		Market Model		Moderating Effect
	Std. Est	P (>|z|)	Std. Est	P (>|z|)	
boc <—cp	0.048	0.4	0.11	0.08	0.062
boc <—psm	**0.296**	**0****	**0.474**	**0****	0.178
boc <—si	**0.236**	**0.004****	-0.111	0.086	0.125
boc <—sob	**0.178**	**0.017***	0.107	0.088	0.071
brc <—cp	-0.058	0.376	-0.081	0.235	0.023
brc <—psm	0.121	0.17	**0.388**	**0****	0.267
brc <—si	**0.226**	**0.016***	-0.114	0.1	0.112
brc <—sob	0.146	0.089	0.125	0.064	0.021
lic <—cp	**0.373**	**0****	**0.364**	**0****	0.009
lic <—psm	**-0.235**	**0.008****	**-0.125**	**0.033***	0.110
lic <—si	0.182	0.053	**0.349**	**0****	0.167
lic <—sob	0.002	0.985	0.067	0.262	0.065
so <—boc	**0.170**	**0.027***	**0.188**	**0.008****	0.018
so <—brc	0.024	0.721	**0.162**	**0.019***	0.138
so <—lic	0.056	0.390	0.121	0.104	0.065
so <—cp	**0.154**	**0.017***	**0.311**	**0****	0.157
so <—psm	0.101	0.236	-0.081	0.295	0.02
so <—si	0.107	0.229	-0.018	0.808	0.089
so <—sob	**0.235**	**0.003****	0.034	0.615	0.201

TABLE 4.8
Results of Hypothesis Testing

Hypothesis Description	Result
Bonding Capital has a significant relationship with Social Outcome	Supported for both Models
Bridging Capital has a significant relationship with Social Outcome	Not Supported for Funded Model
	Supported for Market Model
Linking Capital has a significant relationship with Social Outcome	Not Supported for both Models
Sense of Belongingness has a significant relationship with Bonding Capital	Supported for Funded Model
	Not Supported for Market Model
Sense of Belongingness has a significant relationship with Bridging Capital	Not Supported for both Models
Sense of Belongingness has a significant relationship with Linking Capital	Not Supported for both Models
Social Identity has a significant relationship with Bonding Capital	Supported for Funded Model
	Not Supported for Market Model
Social Identity has a significant relationship with Bridging Capital	Supported for Funded Model
	Not Supported for Market Model
Social Identity has a significant relationship with Linking Capital	Not Supported for Funded Model
	Supported for Market Model
Community Participation has a significant relationship with Bonding Capital	Not Supported for both Models
Community Participation has a significant relationship with Bridging Capital	Not Supported for both Models
Community Participation has a significant relationship with Linking Capital	Supported for both Models
Pro-Social Motivation has a significant relationship with Bonding Capital	Supported for both Models
Pro-Social Motivation has a significant relationship with Bridging Capital	Not Supported for Funded Model
	Supported for Market Model
Pro-Social Motivation has a significant relationship with Linking Capital	Supported for both Models

bridging capital to social outcome were significant (p-values of 0.008, and 0.019, respectively). Linking capital was found to have no significant impact in creating social outcome for this type of projects.

This study identified pro-social motivation of stakeholders as the main antecedent to creation of social capital in social enterprises. The sense of belongingness also came out as another important factor. These two findings together discuss the need for stakeholders of a social enterprise to have close bondages with immediate community in general and to the social enterprise in particular to create the intended social outcome. Hence, this study also highlights the need for creation of an environment with trust and mutual respect within and outside of a social enterprise. Also, the social identity of actors involved in a social enterprise was found to be a significant predictor of social capital. It was also evident that social enterprises which foster a strong sense of identity would be more effective in creating social capital which in turn will create social outcome.

The study also found that community participation impacted social outcome directly than mediating through creation of any type of social capital. This result suggests that, social enterprises designed to develop social interactions would create better, scalable, and tangible outcomes. This finding was evidently reflected in both funded and market models of social entrepreneurship. It was also found that community participation led to more linking capital for both models of social entrepreneurship. It was found that social outcome was created distinctly different in funded and market models of social entrepreneurship. While funded model created the social capital, the market model utilized social capital to create social outcomes.

The above results and observations conclude that, while the funded type of social project would lead to more creation of social capital, the market type would lead to leverage on social capital in crating favorable social outcomes.

4.6   CONCLUSION

The study tries to understand the process of social capital creation and subsequent creation of outcomes of social enterprises. Social outcome was the dependent variable. Social capital was the intervening variable and pro-social motivation, social identity, community participation, and sense of belongingness were the independent variables. A detailed analysis shows that all three kinds of social capital viz. bonding capital, bridging capital, and linking capital had significant relationships with the social outcome. Pro-social motivation had significant relationship with all three kinds of social capital. However, it was found that pro-social motivation had a negative significant relationship with bridging capital. Community participation had significant relationship with bridging and linking capital. Also it was noted that community participation had a negative significant relationship with bridging capital. However, community participation had no significant relationship with bonding capital. Similarly, sense of belongingness was found to have significant relationship with bonding and bridging capital and its relationship with linking capital was found insignificant. Social identity was found to have a significant relationship with only linking capital while its relationships with bonding as well as bridging capitals were found to be insignificant.

The moderation study of the conceptual framework revealed that different models of social entrepreneurship create social capital and social outcomes differently. While the funded model of social entrepreneurship would focus more on the creation of social capital, the market model would leverage the existing social capital for the creation of social outcomes. This led to the conclusion that a two-phase approach would be more meaningful in implementing social projects—initially as a funded model and later transforming into a self-sustainable market model of social entrepreneurship.

The mediation analysis involving social capital revealed that there is a strong mediation effect of social capital in the process of social outcome creation in market model. Whereas, in funded model, the mediation effect was partial in nature. Hence, the market model would be able to leverage the existing social capital better for the creation of social outcome. In the funded model of social entrepreneurship, creation of both social capital and outcomes happen simultaneously.

4.7 LIMITATIONS & SCOPE FOR FUTURE RESEARCH

This study was restricted to only four antecedent variables of social capital. There could be other relevant antecedents too. A study covering those variables could also be possible in the same context, which would add further clarity to the theoretical understanding of the construct of social capital. The current study is a cross-sectional one. Social capital also undergoes dynamic changes with the passage of time. Hence, exploring the dynamic effects on social capital through a longitudinal study would possible bridge some of the limitations of this research. This kind of study would bring more clarity on the aspect of how time affects the potential of social entrepreneurship in creating social capital. This study provides proof to the argument that the creation of social capital is susceptible for the moderation effect due to the model of social entrepreneurship. This study took only two models viz. funded model and market model, which were at the two ends of the spectrum of financial sustainability of social enterprises. There are other models too like movement model, grass-root model etc. that are distinct in many ways. Further research incorporating the moderation effects of these models of social entrepreneurship could also add to the existing knowledge base.

REFERENCES

Algan, Y. and Cahuc, P. (2010). Inherited trust and growth. *American Economic Review*, 100, 2060–2092.

Anderson, J. C. and Gerbing, D. W. (1988). Structural equation modeling in practice: A review and recommended two-step approach. *Psychological Bulletin*, 103(3), 411–423.

Aydinli, A., Bender, M., Chasiotis, A., Van de Vijver, F. J. and Cemalcilar, Z. (2015). Implicit and explicit prosocial motivation as antecedents of volunteering: The moderating role of parenthood. *Personality and Individual Differences*, 74, 127–132.

Barnes-Mauthe, M., Allen, S., Arita, S., Lynham, J. and Leung, P. (2015). What determines social capital in a social—ecological system? Insights from a network perspective. *Environment Management*, 55, 392–410.

Bentler, P. M. and Bonett, D. G. (1980). Significance tests and goodness of fit in the analysis of covariance structures. *Psychological Bulletin*, 88(3), 588–606.

Bernard, L., Mäs, S., Müller, M., Henzen, C. and Brauner, J. (2014). Scientific geodata infra-structures: challenges, approaches and directions. *International Journal of Digital Earth*, 7(7), 613–633.

Bourdieu, P. (1986). The forms of capital. In J. G. Richardson (Ed.), *Handbook of theory and research for the sociology of education*. Greenwood Press, 241–258.

Byrne, B. M. (2010). *Structural equation modeling with AMOS: Basic concepts, applications and programming* (2nd ed.). Routledge.

Campos, A. C. V., Borges, C. M., Vargas, A. M. D., Gomes, V. E., Lucas, S. D. and Ferreira, E. F. (2015). Measuring social capital through multivariate analyses for the IQ-SC. *BMC Research Notes*, 8(11).

Chazdon, S., Allen, R., Horntvedt, J. and Scheffert, D. R. (2013). *Developing and validating University of Minnesota extension's social capital model and survey*. University of Minnesota.

Chen, H., Meng, T., Coleman, J., Putnam, R., Ferlander, S., Agampodi, T., . . . Kawachi, I. (2015). Bonding, bridging, and linking social capital and self-rated health among chinese adults: use of the anchoring vignettes technique. *PLos One*, 10(11).

Cheung, C.-K. and Leung, K.-K. (2011). Neighborhood homogeneity and cohesion in sustainable community development. *Habitat International*, 35, 564–572.

Chiu, C.-M., Hsu, M.-H. and Wang, E. T. G. (2006). Understanding knowledge sharing in virtual communities: An integration of social capital and social cognitive theories. *Decision Support Systems*, 42, 1872–1888.

Coffé, H. (2009). Social capital and community heterogeneity. *Social Indicators Research*, 91, 155–170.

Coleman, J. (1990). *Foundations of social theory*. Harvard University Press.

Darcy, S., Maxwell, H., Edwards, M., Onyx, J. and Sherker, S. (2014). More than a sport and volunteer organisation: Investigating social capital development in a sporting organisation. *Sport Management Review*, 17, 395–406.

Doh, S. and McNeely, C. L. (2012). A multi-dimensional perspective on social capital and economic development: An exploratory analysis. *Annals of Regional Science*, 49, 821–843.

George, D. and Mallery, P. (2010). *SPSS for Windows step by step: A simple guide and reference, 17.0 update* (10th ed.). Allyn & Bacon.

Grant, A. M. and Sumanth, J. J. (2009). Mission possible? The performance of prosocially motivated employees depends on manager trustworthiness. *Journal of Applied Psychology*, 94(4), 927–944.

Hanifan, L. J. (1916). The rural school community centre. *Annals of the American Academy of Political and Social Sciences*, 67, 130–138.

Harper, R. (2001). *Social capital a review of the literature*. Social Analysis and Reporting Division Office for National Statistics.

Hays, R. A. (2015). Neighborhood networks, social capital, and political participation: The relationships revisited. *Journal of Urban Affairs*, 37(2), 122–143.

Henny, V. L. and Marco, O. (2013). The more trust, the fewer transaction costs - searching for a new management. In Proceedings of the European Conference on Management, Leadership & Governance.

Jaafar, M., Rasoolimanesh, S. M. and Ismail, S. (2017). Perceived sociocultural impacts of tourism and community participation: A case study of Langkawi Island. *Tourism and Hospitality Research*, 17(2), 123–134.

Katherine, A., Thomas, M. R. and Julie, O. A. (2010). Community gardening, neighborhood meetings, and social capital. *Journal of Community Psychology*, 38(4), 497–514.

Kibler, E., Wincent, J., Kautonen, T., Cacciotti, G. and Obschonka, M. (2019). Can prosocial motivation harm entrepreneurs' subjective well-being? *Journal of Business Venturing*, 34(4), 608–624.

Kitchin, P. J. and David Howe, P. (2013). How can the social theory of Pierre Bourdieu assist sport management research? *Sport Management Review*, 16, 123–134.

Krause, D. R., Handfield, R. B. and Tyler, B. B. (2007). The relationships between supplier development, commitment, social capital accumulation and performance improvement. *Journal of Operations Management*, 25, 528–545.

Kulangara, N. P., Jackson, S. A. and Prater, E. (2016). Examining the impact of socialization and information sharing and the mediating effect of trust on innovation capability. *International Journal of Operations & Production Management*, 36(11), 1601–1624.

Lawson, B., Tyler, B. B. and Cousins, P. D. (2008). Antecedents and consequences of social capital on buyer performance improvement. *Journal of Operations Management*, 26, 446–460.

Li, P. P. (2007). Social tie, social capital, and social behaviour: Toward an integrative model of informal exchange. *Asia Pacific Journal of Management*, 24, 227–246.

Li, Y., Ye, F. and Sheu, C. (2014). Social capital, information sharing and performance Evidence from China. *International Journal of Operations & Production Management*, 34(11), 1440–1462.

Nahapiet, J. and Ghoshal, S. (1998). Social capital, intellectual capital, and the organizational advantage. *The Academy of Management Review*, 23(2), 242–266.

Narayan, D. and Cassidy, M. F. (2001). A dimensional approach to measuring social capital: Development and validation of a social capital inventory. *Current Sociology*, 49(2), 59–102.

Neal, J. W. and Neal, Z. P. (2019). Implementation capital: merging frameworks of implementation outcomes and social capital to support the use of evidence-based practices. *Implementation Science*, 14(1), 16.

Nichols, M. F. and Fernandez, M. (2007). Bridging and bonding capital: Pluralist ethnic relations in Silicon Valley. *International Journal of Sociology and Social Policy*, 22(5), 104–122.

Nicholson, S., A Cleland, B. J. and Nicholson snicholson, S. (2016). "'It's making contacts'": notions of social capital and implications for widening access to medical education. *Advances in Health Sciences Education*, 1–14.

Oh, H. (2000). Ties that link: The antecedents of social capital and itseffects on entrepreneurial success. Thesis, Pen State University.

Park, H., Tsusaka, T., Pede, V. and Kim, K.-M. (2017). The impact of a local development project on social capital: Evidence from the Bohol Irrigation Scheme in the Philippines. *Water*, 9(3), 202.

Pastoriza, D., Ariño, M. A. and Ricart, J. E. (2007). Ethical managerial behaviour as an antecedent of organizational social capital. *Journal of Business Ethics*, 78(3), 329–341.

Paul, N., Sherry, K., Jackson, A. and Prater, E. (2016). Examining the impact of socialization and information sharing and the mediating effect of trust on innovation capability. *International Journal of Operations & Production Management*, 36(11), 1601–1624.

Pinho, J. C. (2013). The e-SOCAPIT scale: a multi-item instrument for measuring online social capital. *Journal of Research in Interactive Marketing*, 7(3), 216–235.

Presutti, M. and Boari, C. (2008). Space-related antecedents of social capital: Some empirical inquiries about the creation of new firms. *International Entrepreneurship and Management Journal*, 4(2), 217–234.

Putnam, R. D. (1995). Bowling alone: America's declining social capital. *Journal of Democracy*, 6, 65–78.

Putnam, R. D. (2000). *Bowling alone the collapse and revival of American community.* Simon and Schuster.

Rasoolimanesh, S. M., Jaafar, M., Ahmad, A. G. and Barghi, R. (2017). Community participation in World Heritage Site conservation and tourism development. *Tourism Management*, 58, 142–153.

Rasoolimanesh, S. M., Ringle, C. M., Jaafar, M. and Ramayah, T. (2017). Urban vs. rural destinations: Residents' perceptions, community participation and support for tourism development. *Tourism Management*, 60, 147–158.

Richey, S. (2007). Manufacturing trust: Community currencies and the creation of social capital. *Political Behavior*, (29), 69–88.

Robin, L. J., Patrick, J. S. and Natasha, D. W. (2005). Developing social capital through participation in organized youth programs. *Journal of Community Psychology*, 33(1), 41–55.

Rydin, Y. and Holman, N. (2004). Re-evaluating the contribution of social capital in achieving sustainable development. *Local Environment*, 9(2), 117–133.

Sabatini, F. (2009). Social capital as social networks: A new framework for measurement and an empirical analysis of its determinants and consequences. *The Journal of Socio-Economics*, 38(3), 429–442.

Santos, F., Pache, A.-C. and Birkholz, C. (2015). Making hybrids work: Aligning business models and organizational design for social enterprises. *California Management Review*, 57(3), 36–58.

Seferiadis, A. A., Cummings, S., Zweekhorst, M. B. M. and Bunders, J. F. G. (2015). Producing social capital as a development strategy: Implications at the micro-level. *Progress in Development Studies*, 15(2), 170–185.

Shoji, M. N., Haskins, A. R., Rangel, D. E. and Sorensen, K. N. (2014). The emergence of social capital in low-income Latino elementary schools. *Early Childhood Research Quarterly*, (29), 600–613.

Sperber, A. D. (2004). Translation and validation of study instruments for cross-cultural research. *Gastroenterology*, 126(1), 124–128.

Stuart, E. (2013). Virtual community participation and motivation: Cross-disciplinary theories. *Online Information Review*, 37(1), 154–155.

Subramanian, S. V., Lochner, K. A. and Kawachi, I. (2003). Neighborhood differences in social capital: A compositional artifact or a contextual construct? *Health and Place*, 9, 33–44.

Svendsen, G. and Srensen, J. F. L. (2006). Socioeconomic power of social capital the socioeconomic power of social capital a double test of Putnam's civic society argument. *International Journal of Sociology and Social Policy*, 2610(9), 411–429.

Tonts, M. (2005). Competitive sport and social capital in rural Australia. *Journal of Rural Studies*, 21(2), 137–149.

Urban, B. (2015). An exploratory study on outcomes of social enterprises in South Africa. *Journal of Enterprising Culture*, 23(2), 271–297.

Villalonga-Olives, E., Adams, I. and Kawachi, I. (2016). The development of a bridging social capital questionnaire for use in population health research. *SSM—Population Health*, 2, 613–622.

Werts, C. E., Rock, D. R., Linn, R. L. and Jöreskog, K. G. (1978). A general method of estimating the reliability of a composite. *Educational and Psychological Measurement*, 38(4), 933–938.

Woolcock, M. and Narayan, D. (2000). Social capital: Implications for development theory, research, and policy. *The World Bank Research Observer*, 15(2), 225–249.

Yamaoka, K., Tomosho, T., Mizoguchi, M. and Sugiura, M. (2008). Social capital accumulation through public policy systems implementing paddy irrigation and rural development projects. *Paddy and Water Environment*, 6, 115–128.

Yim, B. and Leem, B. (2013). The effect of the supply chain social capital. *Industrial Management & Data Systems*, 113(3), 324–349.

Zetter, R., Griffiths, D. and Sigona, N. (2006). Integrative paradigms, marginal reality: Refugee community organisations and dispersal in Britain. *Journal of Ethnic and Migration Studies*, 32(5), 881–898.

5 Study on Data-Driven Decision-Making in Entrepreneurship

Vimlesh Kumar Ojha, Sanjeev Goyal and Mahesh Chand

CONTENTS

5.1 Introduction 75
5.2 Literature Review 77
    5.2.1 Traditional Approaches of Decision-Making 77
    5.2.2 Data Driven Decision-Making in Industry 4.0 79
5.3 Conclusion & Discussion 82
References 83

5.1 INTRODUCTION

Entrepreneurship creates a competitive edge through enhancing organizational capabilities, making strategic decisions, and using innovation to achieve final clients (Djordjevic, 2013). Lower costs of production, huge production volume, positive cash flow, improved quality, lower wastages, and customer satisfaction are the some key factors through which the competitive advantage and entrepreneurial factor impacts on organizational capacities may be measured (Micheels and Gow, 2008). Making a decision involves all the existing choices being taken into consideration, and then, on the basis of the goals, objectives, and desires of the decision-maker, the best alternative is chosen (Connors et al., 2015; Jain and Chand, 2021). As discussed in Figure 5.1, Entrepreneurial decision-making involves seven decision-making activities: opportunity assessment decision, entrepreneurial entry decision, decision about exploiting opportunities, entrepreneurial exit decision, heuristics and biases in decision-making process, Entrepreneurial decision-maker characteristics and decision-making environment (Shepherd et al., 2015). Due to rapid technological and financial progress in the last fifty years, entrepreneurs have witnessed a huge change. Now, entrepreneurs have to solve such complex decision-making problems that involve many criteria, sub-criteria, and alternatives. Researchers have discussed various styles of decision-making on entrepreneurship, among these most cognitive evaluations of entrepreneurial activity are based on a logical approach (Nagar et al., 2021). The logical approach takes into account the matters like real options

DOI: 10.1201/9781003256663-5

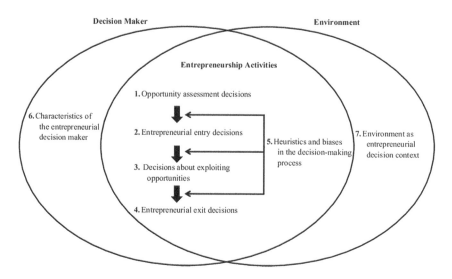

FIGURE 5.1   Entrepreneurial Decision-Making.

variables, risk-taking proclivity, up-to-date information etc. (Ireland et al., 2003). Business intelligence assists in the acquisition of essential data from a variety of unstructured sources, as well as the translation of that data into useable information that enables entrepreneurs to make better policy choices and boost efficiency and productivity (Chand, 2021). A parallel development in mathematical science is ensuring the decision-making easier by formulizing the problems based upon the past data which results transparency into all aspects of decision (Khaira and Dwivedi, 2018). Globalization has increased the competition in production rapidly, which forces entrepreneurs to produce a variety of products and quickly respond to the changes in the market. The rapidly changing technology in the product sector has necessitated a correspondingly quick response from entrepreneurs (Singh et al., 2019). Because of increased competitiveness, rapid technological innovation, and rapid changes in customer demand, entrepreneurs running manufacturing organizations recognized the significance of advanced manufacturing systems with a wide coverage of practical abilities (Gunasekaran, 1999). These goals can be achieved by using advanced manufacturing systems (AMS), which combine robotics, computer numerical control machines (CNC), and automated material handling. The key benefits of FMS are more product variety, greater quality, and lower waste and setup costs (Karsak and Kuzgunkaya, 2002; Sindhwani et al., 2021). The development of an advanced production facility popularly known as Industry 4.0 which integrates data science and IoT with machining centers (NC, CNC and DNC), flexible manufacturing systems, robotic system, and automated quality control and storage system is currently the prime focus of the production organizations (Chakraborty, 2011; Sindhwani et al., 2022a). Since Industry 4.0 offers an automated production system, but to fulfill the expectations of rapid changing demand, the selection of suitable strategies of production, component design, stages of production, work pieces and tools, machines tools is still a challenging and important activity. Making choices is tough, as making decisions is a chief challenge today (Esmaeilian et al., 2016; Sindhwani et al., 2022b). Understanding current and future developments, taking all factors affecting the entire production environment into consideration, accessing a variety of methods and techniques available for decision-making, measuring the characteristics of processes used for decision-making, and finally, manufacturing properly depending upon numerous issues related to each stage like design, quality, planning, and maintenance are the processes of decision-making in a manufacturing organization (Kiker et al., 2005; Singh et al., 2022). When it comes to business knowledge and decision-making, every organization faces difficulties such as plan failure, lack of readiness, resource unavailability, and risk-taking capacities. This study examines different classic decision-making strategies as well as new data-driven decision-making techniques that aid entrepreneurs in making better decisions and reducing plan failure and risk-taking abilities.

5.2 LITERATURE REVIEW

To get success in an increasingly complicated environment, entrepreneurs must have the ability to perceive the larger picture while also considering the finer details. To do so, they must be able to see whole concepts as particular elements while making any decision. Because of this, the quest for models and approaches to aid decision-making managers is not confined to any one style of management, nor is it restricted to manufacturing and remodeling (Svensson, 2003). Generally, decision-making techniques are broadly categorized into two categories: traditional approaches to decision-making and data-driven approaches to decision-making. These two decision-making approaches are discussed in this section.

5.2.1 TRADITIONAL APPROACHES OF DECISION-MAKING

Several multi-objective decision-making (MODM) methodologies are currently available for manufacturing decision-making. For solving the selection problems through MODM methods, each alternative on the set of objectives must be considered in order to make a decision. (Sindhwani et al., 2021). Multi-Criteria Decision Analysis (MCDA), Multi-Dimensions Decision-Making (MDDM), and Multi-Attributes Decision-Making (MADM) are all terms used to describe MODM (Zardari et al., 2014). Many decision-making approaches are frequently utilized to assist in identifying the best option from a fixed group of options. Many decision-making techniques (MADM) have been effectively used in the selection of materials (Davim and Reis, 2005), thermal power plants (Garg et al., 2007; Sindhwani et al., 2021a) industrial robots, project testing, cell phones, manufacturing, systems flexible production, performance measurement models for production organizations, plant structure, and other areas.

Manufacturing decision-makers are frequently faced with the task of considering a large number of options and then picking one based on competing principles. To aid those decisions in the modern manufacturing environment, Vinet and Zhedanov

(2011) offered two approaches: graph theory and matrix method, in addition to fuzzy multiple attribute decisions. Karsak and Kuzgunkaya (2002) suggested a programming approach based on fuzzy multiple objective for the selection of best suitable among various options of flexible manufacturing system (FMS). According to Byun and Lee (2005), MADM gives an effective framework for evaluating RP systems based on their ratings for several criteria, as well as a procedure for determining the most suited RP system for the component's eventual function. Yurdakul and Çoğun (2003) and Sindhwani et al. (2021b) devised a multidisciplinary method for identifying and classifying relevant NTMPs in a specific system. Agent-based modelling (ABM) has been proved to be an ideal foundation for the building of robust and accurate "what-if" circumstances inside production and operational systems (Nilsson and Darley, 2006). Multi-objective optimization on the basis of ratio analysis (MOORA) is a relatively new MODM method which resolves various decision-making problems that are commonly faced in real-time production environments (Chakraborty, 2011). The act of concurrently increasing more than one contending attributes according to some fixed restrictions is known as multi-objective optimization (or programming) (Achebo and Odinikuku, 2015). Some of the most commonly used decision-making methods are below as:

A. **Analytic Hierarchy Process (AHP):** Due to its simplicity and versatility, AHP is one of the most well-known MCDM methods around the globe by researchers for decision-making, mainly where the many criteria are involved (Khaira and Dwivedi, 2018). To help the analysts identify the desired opportunity, an analytic hierarchy process (AHP) is used to reduce complex decision issues methodically and analytically (Wang et al., 2017). In a typical AHP application, first the problem statement is defined and then we identify the goal which has to be achieved, which is followed by pairwise comparison of components with reference to criteria's then a hierarchy consisting a family tree was structured which is viewed and structured form in demonstrating a problem (Srdjevic and Medeiros, 2008).

B. **Analytic Network Process (ANP):** ANP, just like AHP, uses pairwise comparisons for measuring the intangible factors and describes the dominance of one factor over another factor in relation to the shared objective (Chung et al., 2005). The application of the ANP is found in various literature like: in semiconductor fabricator for constitution of product mix planning (Chung et al., 2005); for development of a system for selection of employees on the weightage of different selection factors (Görener, 2012); in the food industry for choosing the best quality management system (Radaev and Bochkov, 2009), for analysis of select issues of green supply chain management (GSCM) (Chand et al., 2018).

C. **Interpretive structural modeling (ISM):** Interpretive structural modelling (ISM) is a strategy for detecting linkages between particular things that constitute a problem or an issue (Singh and Kant, 2008). Mehregan et al. (2014) used ISM to analyze significant vendor selection factors and show interrelationships. Kannan and Haq (2007) and Kumar et al. (2021) deployed ISM in the analysis of criteria and sub-criteria for the selection of a supplier for built-in-order supply chains. Chand et al. (2019) applied an ISM approach for analysis of lean practices in manufacturing industries. Mathiyazhagan et al. (2013) adopted ISM to identify and analyze green supply chain management implementation hurdles.

D. **TOPSIS**: TOPSIS is used to weigh the alternatives while taking into account the distance to the ideal solution and the negative-ideal solution for each alternative, and picking the one that is the closest to the ideal solution as the best option (Opricovic and Tzeng, 2004). Many writers proposed a combined strategy in which TOPSIS is used in conjunction with AHP to ensure that a MCDM process is comprehensive, taking into account both subjective and objective factors.

E. **PROMETHEE:** PROMETHEE stands for Preference Ranking Organization Method for Enrichment Evaluations. It comes under the category of outranking methods and was introduced by Brans et al. (1984). To establish partial binary relations between alternatives a1 and a2, PROMETHEE, like other outranking algorithms, compares them pairwise inside each single criteria (Uzun et al., 2021). Particularly in manufacturing, some applications of this techniques are as, scheduling (Belz and Mertens, 1996), maintenance planning (Kralj and Petrovic, 1995), equipment selection (Dağdeviren, 2008) and manufacturing system/technology selection (Gurumurthy and Kodali, 2008).

F. **Structural equation modelling:** Multivariate statistical analysis techniques such as structural equation modelling (SEM) are often used for the investigation of structural connections (Buhi et al., 2007; Kumar et al., 2016). As a combination of multiple regression and component analysis, it is often used to examine the structural correlations that tend to exist between latent constructs and the variables that are used to measure these latent constructs (Nunkoo and Ramkissoon, 2012).

5.2.2   DATA DRIVEN DECISION-MAKING IN INDUSTRY 4.0

Literature on innovation and entrepreneurship has mostly identified a dependable and static entrepreneurial process for assessing a new idea that supports a market opportunity. Adaptation of data-driven technology, the entrepreneurial process has become less constrained on product scope, market research and boundaries of entrepreneurial activities (Nambisan, 2017). The constituents of Industry 4.0 like sensors, automated machines, smart devices generate oceans of data continuously. The manufacturing sector nowadays is undergoing a not ever seen surge in data availability (Thoben et al., 2017). These data originate from sensors fitted on the production lines, machine tools parameters, environmental data and customer feedback data (Mittal et al., 2019) and this rise of the availability of massive volumes of data is described as Big data (J. Lee et al., 2013). There is a huge potential to improve product quality and production capacity by utilizing this largely available data for the decision-making (Elangovan et al., 2015). In the recent years, extensive improvements were witnessed in the field of data storage and processing and thus it becomes easier now to analyze the data obtained from machine more rapidly and profoundly. These advancements have boosted the value of Artificial Intelligence (AI) technology and ushered in a new stage of development known as Industry 4.0. In the different parts of the world, this movement is known by a different nomenclature like in Germany it is called Industry 4.0, in US it is known as Smart Manufacturing, and people in South Korea calls it Smart Factory. Internet of Things (IoT), big data analysis, processing of complex work, production and sales forecasting, service-oriented architectures (SoA), autonomous units, adaptive and predictive control, and other terms arose from the fourth industrial revolution (Industry 4.0). More complexities of the smart manufacturing systems open the path for new innovation and development activities. To use such novel systems in an effectively and efficiently, new approaches for design, optimization, control, observation, and monitoring are needed. In smart manufacturing systems, various kind of data like data from networked systems and data from production lines are blended. Using the vast amount of data created by the systems to make efficient decisions that can boost production, reduce waste, and improve quality is one of the main difficulties in Industry 4.0 (Porter and Heppelmann, 2015). These complex decision-making problems cannot be done naturally and blindly. The decision-making model supported by the past data should be used for making such complex decisions. Such type of decision-making models which are supported by the big data is often called data driven decision-making which involves six phases (Table 5.1):

A lot of work has been done on the application of data science in the Industry 4.0 for making decisions. As shown in Figure 5.2, the data driven approach has been applied in almost all contexts, which is closely related to the process of manufacturing organization. Yu and Matta (2016) developed a data-driven strategy to detect machine bottleneckness using multiple arithmetical techniques to reduce false detection rates. Omar et al. (2017) developed a data driven model and discussed a methodology for estimation of system data form different points of work in process and used this estimate to discharge production.

Industry 4.0 which is generally known as the fourth generation of manufacturing employs the ideas like cyber physical system, augmented and virtual reality, data

TABLE 5.1
Phases in Data Driven Decision-Making

Phase No.	Phase Name	Major Actions in the Phase
I	Understanding the objective	Objectives identification, assessment of situation, goals determination.
II	Understanding of data	Raw data collection, data exploration and its description, authentication of data quality.
III	Preparation of data	Data pre-processing, profiling, cleansing, validation and transformation
IV	Modelling of data	Finding the best technique of modelling, test design generation, model building.
V	Evaluation of data	Calculation of final result after analysis, outcome optimization and forecasting the next steps.
VI	Deployment of result	Exploitation of lessons learned from result, further interpretations.

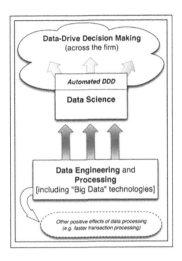

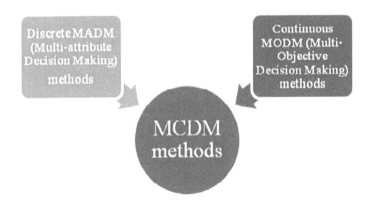

FIGURE 5.2   Data driven decision-making in context of closely related process of organization.

driven decision-making to develop a smart factory (Bottani et al., 2017). Connection, cyber, cloud, content, customization and community are the six Cs which describes the significance of big data in Industry 4.0 (Miragliotta et al., 2018). Zhong et al. (2017) suggested an Internet of Things (IoT) and RFID-enabled intelligent manufacturing shop floor framework. Their platform was able to collect vast amounts of data for analysis and management consequences. In order to establish casual links in the development of a reverse engineering method for mining data is developed by (Pmb,

2016). For long considered as one of the most essential facilitators of process and quality control, big data is now becoming even more crucial (Wang et al., 2017). Forza and Salvador (2008) and Shanker et al. (2019) addressed the necessity to use and integrate data science in product design and development, manufacturing and quality control and customer feedback management systems to improve overall decision-making. Various researchers have applied data driven approaches in all stages of manufacturing from starting from the design to the production as summarized below:

A technique of enabling a computer, robot, or other machine to think, process information, and act on its own is known as artificial intelligence, or AI for short. In other words, providing robots the capacity to think like humans is known as artificial general intelligence (AGI) (Huang and Rust, 2018). Bullers et al. (1980) looked at common difficulties in planning and control of automated manufacturing system, and described the applications of artificial intelligence approaches in solving such problems. Chien et al. (2020) found that additive manufacturing and data-driven maintenance are critical to the growth of Smart Factories. Li et al. (2017) examined the fast growth of fundamental technologies in the modern age of "Internet plus AI," which is producing a substantial change in the manufacturing industry's models, methods, and ecosystems, as well as some suggestions for the use of AI in intelligent manufacturing in China. Machine learning and artificial intelligence (AI) technologies play a critical role in contemporary production, especially in the environment of the Industry 4.0 (Zeba et al., 2021). Due to vast volume and large variety data generated in industries (Jin et al., 2021) and the enhanced usability and capability of accessible Machine Learning (ML) tools, the use of machine learning techniques has expanded in the recent period (Pham and Afify, 2005). Today, machine learning is widely used in several aspects of production, such as optimization, control, and troubleshooting (Pham and Afify, 2005) (Table 5.2). The advantages of the data driven decision-making over conventional MCDM techniques are shown in Table 5.3.

5.3   CONCLUSION & DISCUSSION

In the context of Industry 4.0, the availability of a large amount of data generated by smart manufacturing is challenging existing decision-making approaches. Data-driven decision-making aids entrepreneurs in making competitive price decisions, expanding into new areas, and staying up to date with customer behavior changes.

TABLE 5.2
Applications of Data Driven Decision-Making in Manufacturing

Different Process of Organization	Literature having Data Driven Approach
Process and quality monitoring	(Qian et al., 2021); (Lee and Tsai, 2019).
Process capability	(Lin et al., 2019).
Monitoring tools	(Liu and Jiang, 2016); (W. Liu et al., 2020).
Product development and design	(Leng et al., 2020);(He et al., 2015).
Supply chain management	(Chae et al., 2014); (Barbosa et al., 2018); (W. Yu et al., 2019).
Customer research	(Abdolvand et al., 2015); (Papinniemi et al., 2014).
Online platforms and social media	(Chien et al., 2020); (Chong et al., 2017); (Jiang et al., 2017).

TABLE 5.3

Advantages of DDDM over Conventional Decision-Making Techniques

Parameters for Differentiation	Conventional MCDM Techniques	Data Driven Decision-Making (DDDM)
Approach type	MCDM techniques are mostly survey-based approach which always have the problem of transparency and accountability.	It is a data-based approach which leads to greater transparency and accountability.
Concepts	Hypothetical concepts.	Demonstrable and analytical concepts.
Reliability considerations	Decisions heavily rely on the mood of the experts so chances that social, mental, and other factors can impact decisions.	Decisions totally rely on the data and technology so very few chances that decision will have any impact of surroundings.
Suitability	Not suitable for the continuous improvement.	Suitable for continuous improvement.
Linking the decisions with business processes	No provisions to link business decision with analytics.	It ties business decisions to analytics inside.
Feedback support mechanism	No feedback for market research.	Provides clear feedback for market research.
Consistency in results	Lesser consistency, productivity, and efficiency.	It enhances consistency, productivity and efficiency.

Based upon the reviewed articles, it is seen that the entrepreneurs of the manufacturing sector are going to see tremendous change in the near future. The data generated at each stage of the manufacturing like design, production, consumption, and decomposing, can be largely utilized in the near future, which will change the traditional decision-making approach to a data-driven decision-making one. There is strong evidence that entrepreneurs can improve industrial performance substantially via data-driven decision-making methods based on big data. Data science supported data-driven decision-making will allow entrepreneurs to make decisions automatically at a massive scale. The study advocates the utility of Data-Driven Decision-Making approach as reliable tool to unearth new facts and solve the most difficult challenges for present day entrepreneurial setups.

REFERENCES

Abdolvand, N., Albadvi, A. and Aghdasi, M. (2015). Performance management using a value-based customer-centered model. *International Journal of Production Research*, 53(18), 5472–5483.

Achebo, J. and Odinikuku, W. E. (2015). Optimization of gas metal arc welding process parameters using standard deviation (SDV) and multi-objective optimization on the basis of ratio analysis (MOORA). *Journal of Minerals and Materials Characterization and Engineering*, 3(4), 298–308.

Barbosa, M. W., Vicente, A. de la C., Ladeira, M. B. and de Oliveira, M. P. V. (2018). Managing supply chain resources with Big Data Analytics: a systematic review. *International Journal of Logistics Research and Applications*, 21(3), 177–200.

Belz, R. and Mertens, P. (1996). Combining knowledge-based systems and simulation to solve rescheduling problems. *Decision Support Systems*, 17(2), 141–157.

Bottani, E., Cammardella, A., Murino, T. and Vespoli, S. (2017). From the cyber-physical system to the digital twin: The process development for behaviour modelling of a cyber guided vehicle in M2M logic. Proceedings of the Summer School Francesco Turco, 2017 Septe, 96–102.

Brans, J. P., Mareschal, B. and Vincke, P. (1984). *Promethee: A new family of outranking methods in multicriteria analysis*. ULB—Universite Libre de Bruxelles, 477–490.

Buhi, E. R., Goodson, P. and Neilands, T. B. (2007). Structural equation modeling: A primer for health behavior researchers. *American Journal of Health Behavior*, 31(1), 74–85.

Bullers, W. I., Nof, S. Y. and Whinston, A. B. (1980). Artificial intelligence in manufacturing planning and control. *AIIE Transactions*, 12(4), 351–363.

Byun, H. S. and Lee, K. H. (2005). A decision support system for the selection of a rapid prototyping process using the modified TOPSIS method. *International Journal of Advanced Manufacturing Technology*, 26(11–12), 1338–1347.

Chae, B. K., Olson, D. and Sheu, C. (2014). The impact of supply chain analytics on operational performance: A resource-based view. *International Journal of Production Research*, 52(16), 4695–4710.

Chakraborty, S. (2011). Applications of the MOORA method for decision making in manufacturing environment. *International Journal of Advanced Manufacturing Technology*, 54(9–12), 1155–1166.

Chand, M. (2021). Strategic assessment and mitigation of risks in sustainable manufacturing systems. *Sustainable Operations and Computers*, 2, 206–213.

Chand, M., Bhatia, N. and Singh, R. K. (2018). ANP-MOORA based approach for analysis of select issues of green supply chain management (GSCM). *Benchmarking: An International Journal*, 25(2), 642–659.

Chand, M., Suraj and Mishra, O. P. (2019). Analysis of lean practices in manufacturing industries: An ISM approach. *International Journal of Six Sigma and Competitive Advantage*, 11(1), 73–94.

Chien, W. C., Huang, S. Y., Lai, C. F., Chao, H. C., Hossain, M. S. and Muhammad, G. (2020). Multiple contents offloading mechanism in AI-enabled opportunistic networks. *Computer Communications*, 155, 93–103.

Chong, A. Y. L., Ch'ng, E., Liu, M. J. and Li, B. (2017). Predicting consumer product demands via Big Data: the roles of online promotional marketing and online reviews. *International Journal of Production Research*, 55(17), 5142–5156.

Chung, S. H., Lee, A. H. I. and Pearn, W. L. (2005). Product mix optimization for semiconductor manufacturing based on AHP and ANP analysis. *International Journal of Advanced Manufacturing Technology*, 25(11–12), 1144–1156.

Connors, B. B., Rende, R. and Colton, T. J. (2015). Decision-making style in leaders: Uncovering cognitive motivation using signature movement patterns. *International Journal of Psychological Studies*, 7(2).

Dağdeviren, M. (2008). Decision making in equipment selection: An integrated approach with AHP and PROMETHEE. *Journal of Intelligent Manufacturing*, 19(4), 397–406.

Davim, J. P. and Reis, P. (2005). Damage and dimensional precision on milling carbon fiber-reinforced plastics using design experiments. *Journal of Materials Processing Technology*, 160(2), 160–167.

Djordjevic, B. (2013). Strategic entrepreneurship. *Mediterranean Journal of Social Sciences*, 4(15).

Elangovan, M., Sakthivel, N. R., Saravanamurugan, S., Nair, B. B. and Sugumaran, V. (2015). Machine learning approach to the prediction of surface roughness using statistical features of vibration signal acquired in turning. *Procedia Computer Science*, 50, 282–288.

Esmaeilian, B., Behdad, S. and Wang, B. (2016). The evolution and future of manufacturing: A review. *Journal of Manufacturing Systems*, 39, 79–100.

Forza, C. and Salvador, F. (2008). Application support to product variety management. *International Journal of Production Research*, 46(3), 817–836.

Garg, R. K., Agrawal, V. P. and Gupta, V. K. (2007). Coding, evaluation and selection of thermal power plants—A MADM approach. *International Journal of Electrical Power and Energy Systems*, 29(9), 657–668.

Görener, A. (2012). Comparing AHP and ANP: An application of strategic decisions making in a manufacturing company. *International Journal of Business and Social Science*, 3(11), 194–208.

Gunasekaran, A. (1999). Agile manufacturing: A framework for research and development. *International Journal of Production Economics*, 62(1), 87–105.

Gurumurthy, A. and Kodali, R. (2008). A multi-criteria decision-making model for the justification of lean manufacturing systems. *International Journal of Management Science and Engineering Management*, 3(2), 100–118.

He, W., Tian, K., Xie, X., Wang, X., Li, Y., Wang, X. and Li, Z. (2015). Individualized surgical templates and titanium microplates for le Fort i osteotomy by computer-aided design and computer-aided manufacturing. *Journal of Craniofacial Surgery*, 26(6), 1877–1881.

Huang, M. H. and Rust, R. T. (2018). Artificial intelligence in service. *Journal of Service Research*, 21(2), 155–172.

Ireland, R. D., Hitt, M. A. and Sirmon, D. G. (2003). A model of strategic entrepreneurship: The construct and its dimensions. *Journal of Management*, 29(6), 963–989.

Jain, V. and Chand, M. (2021). Decision making in FMS by COPRAS approach. *International Journal of Business Performance Management*, 22(1), 75.

Jiang, C., Liu, Y., Ding, Y., Liang, K. and Duan, R. (2017). Capturing helpful reviews from social media for product quality improvement: A multi-class classification approach. *International Journal of Production Research*, 55(12), 3528–3541.

Jin, Y., Wang, H. and Sun, C. (2021). Introduction to machine learning. *Studies in Computational Intelligence*, 975(34), 103–145.

Kannan, G. and Haq, A. N. (2007). Analysis of interactions of criteria and sub-criteria for the selection of supplier in the built-in-order supply chain environment. *International Journal of Production Research*, 45(17), 3831–3852.

Karsak, E. E. and Kuzgunkaya, O. (2002). A fuzzy multiple objective programming approach for the selection of a flexible manufacturing system. *International Journal of Production Economics*, 79(2), 101–111.

Khaira, A. and Dwivedi, R. K. (2018). A State of the art review of analytical hierarchy process. *Materials Today: Proceedings*, 5(2), 4029–4035.

Kiker, G. A., Bridges, T. S., Varghese, A., Seager, P. T. P. and Linkov, I. (2005). Application of multicriteria decision analysis in environmental decision making. *Integrated Environmental Assessment and Management*, 1(2), 95–108.

Kralj, B. and Petrovic, R. (1995). A multiobjective optimization approach to thermal generating units maintenance scheduling. *European Journal of Operational Research*, 84(2), 481–493.

Kumar, R., Kumar, V. and Singh, S. (2016). Relationship establishment between lean manufacturing and supply chain characteristics to study the impact on organisational performance using SEM approach. *International Journal of Value Chain Management*, 7(4), 352–367.

Kumar, R., Sindhwani, R., Arora, R. and Singh, P. L. (2021). Developing the structural model for barriers associated with CSR using ISM to help create brand image in the manufacturing industry. *International Journal of Advanced Operations Management*, 13(3), 312–330.

Lee, C. Y. and Tsai, T. L. (2019). Data science framework for variable selection, metrology prediction, and process control in TFT-LCD manufacturing. *Robotics and Computer-Integrated Manufacturing*, 55, 76–87.

Lee, J., Lapira, E., Bagheri, B. and Kao, H. an. (2013). Recent advances and trends in predictive manufacturing systems in big data environment. *Manufacturing Letters*, 1(1), 38–41.

Leng, J., Ruan, G., Jiang, P., Xu, K., Liu, Q., Zhou, X. and Liu, C. (2020). Blockchain-empowered sustainable manufacturing and product lifecycle management in industry 4.0: A survey. *Renewable and Sustainable Energy Reviews*, 132.

Li, B., Hou, B., Yu, W., Lu, X. and Yang, C. (2017). Applications of artificial intelligence in intelligent manufacturing: a review. *Frontiers of Information Technology and Electronic Engineering*, 18(1), 86–96.

Lin, K. P., Yu, C. M. and Chen, K. S. (2019). Production data analysis system using novel process capability indices-based circular economy. *Industrial Management and Data Systems*, 119(8), 1655–1668.

Liu, C. and Jiang, P. (2016). A cyber-physical system architecture in shop floor for intelligent manufacturing. *Procedia CIRP*, 56, 372–377.

Liu, W., Kong, C., Niu, Q., Jiang, J. and Zhou, X. (2020). A method of NC machine tools intelligent monitoring system in smart factories. *Robotics and Computer-Integrated Manufacturing*, 61.

Mathiyazhagan, K., Govindan, K., NoorulHaq, A. and Geng, Y. (2013). An ISM approach for the barrier analysis in implementing green supply chain management. *Journal of Cleaner Production*, 47, 283–297.

Mehregan, M. R., Hashemi, S. H., Karimi, A. and Merikhi, B. (2014). Analysis of interactions among sustainability supplier selection criteria using ISM and fuzzy DEMATEL. *International Journal of Applied Decision Sciences*, 7(3), 270–294.

Micheels, E. T. and Gow, H. R. (2008). Market orientation, innovation and entrepreneurship: An empirical examination of the illinois beef industry. *International Food and Agribusiness Management Review*, 11(3), 31–55.

Miragliotta, G., Sianesi, A., Convertini, E. and Distante, R. (2018). Data driven management in Industry 4.0: a method to measure. *Data Productivity*, 51(11), 19–24.

Mittal, S., Khan, M. A., Romero, D. and Wuest, T. (2019). Smart manufacturing: Characteristics, technologies and enabling factors. *Proceedings of the Institution of Mechanical Engineers, Part B: Journal of Engineering Manufacture*, 233(5), 1342–1361.

Nagar, D., Raghav, S., Bhardwaj, A., Kumar, R., Singh, P. L. and Sindhwani, R. (2021). Machine learning: Best way to sustain the supply chain in the era of industry 4.0. *Materials Today: Proceedings*, 47, 3676–3682.

Nambisan, S. (2017). Digital entrepreneurship: Toward a digital technology perspective of entrepreneurship. *Entrepreneurship: Theory and Practice*, 41(6), 1029–1055.

Nilsson, F. and Darley, V. (2006). On complex adaptive systems and agent-based modelling for improving decision-making in manufacturing and logistics settings: Experiences from a packaging company. *International Journal of Operations and Production Management*, 26(12), 1351–1373.

Nunkoo, R. and Ramkissoon, H. (2012). Structural equation modelling and regression analysis in tourism research. *Current Issues in Tourism*, 15(8), 777–802.

Omar, R. S. M., Venkatadri, U., Diallo, C. and Mrishih, S. (2017). A data-driven approach to multi-product production network planning. *International Journal of Production Research*, 55(23), 7110–7134.

Opricovic, S. and Tzeng, G. H. (2004). Compromise solution by MCDM methods: A comparative analysis of VIKOR and TOPSIS. *European Journal of Operational Research*, 156(2), 445–455.

Papinniemi, J., Hannola, L. and Maletz, M. (2014). Challenges in integrating requirements management with PLM. *International Journal of Production Research*, 52(15), 4412–4423.

Pham, D. T. and Afify, A. A. (2005). Machine-learning techniques and their applications in manufacturing. *Proceedings of the Institution of Mechanical Engineers, Part B: Journal of Engineering Manufacture*, 219(5), 395–412.

Pmb, E. (2016). *2016 multi annual strategic research and innovation agenda for ECSEL joint undertaking MASRIA*. Ecsel.

Porter, M. E. and Heppelmann, J. E. (2015). How smart, connected products are transforming companies. *Harvard Business Review*, 2015.

Qian, F., Jin, Y., Qin, S. J. and Sundmacher, K. (2021). Guest editorial special issue on deep integration of artificial intelligence and data science for process manufacturing. *IEEE Transactions on Neural Networks and Learning Systems*, 32(8), 3294–3295.

Radaev, N. N. and Bochkov, A. V. (2009). Application of the AHP/ANP in food quality management. In Proceedings of ISAHP, 2009.

Shanker, K., Shankar, R. and Sindhwani, R. (2019). Advances in industrial and production engineering. In *Select proceedings of FLAME 2018 book series*. Springer-Nature.

Shepherd, D. A., Williams, T. A. and Patzelt, H. (2015). Thinking about entrepreneurial decision making: Review and research agenda. *Journal of Management*, 41(1), 11–46.

Sindhwani, R., Afridi, S., Kumar, A., Banaitis, A., Luthra, S. and Singh, P. L. (2022b). Can industry 5.0 revolutionize the wave of resilience and social value creation? A multicriteria framework to analyze enablers. *Technology in Society*, 101887.

Sindhwani, R., Hasteer, N., Behl, A., Varshney, A. and Sharma, A. (2022a). Exploring "what," "why" and "how" of resilience in MSME sector: A m-TISM approach. *Benchmarking: An International Journal*, In press (ahead-of-print).

Sindhwani, R., Kumar, R., Behl, A., Singh, P. L., Kumar, A. and Gupta, T. (2021a). Modelling enablers of efficiency and sustainability of healthcare: A m-TISM approach. *Benchmarking: An International Journal*, 29(3), 767–792.

Sindhwani, R., Singh, P. L., Kumar, R., Kumar, B., Mittal, V. K., Dixit, S. and Jindal, A. (2021b). Structural modelling for the factors of facility management in the healthcare industry using the TISM approach. In *Multi-criteria decision modelling*. CRC Press, 1–21.

Singh, M. D. and Kant, R. (2008). Knowledge management barriers: An interpretive structural modeling approach. *International Journal of Management Science and Engineering Management*, 3(2), 141–150.

Singh, P. L., Sindhwani, R., Sharma, B. P., Srivastava, P., Rajpoot, P. and Kumar, R. (2022). Analyse the critical success factor of green manufacturing for achieving sustainability in automotive sector. In *Recent trends in industrial and production engineering*. Springer, 79–94.

Singh, R. K., Kumar, P. and Chand, M. (2019). Evaluation of supply chain coordination index in context to Industry 4.0 environment. *Benchmarking*, 28(5), 1622–1637.

Srdjevic, B. and Medeiros, Y. D. P. (2008). Fuzzy AHP assessment of water management plans. *Water Resources Management*, 22(7), 877–894.

Svensson, L. E. O. (2003). Escaping from a liquidity trap and deflation: The foolproof way and others. *Journal of Economic Perspectives*, 17(4), 145–166.

Thoben, K. D., Wiesner, S. A. and Wuest, T. (2017). "Industrie 4.0" and smart manufacturing-a review of research issues and application examples. *International Journal of Automation Technology*, 11(1), 4–16.

Uzun, B., Almasri, A. and Uzun Ozsahin, D. (2021). Preference ranking organization method for enrichment evaluation (Promethee). *International Journal of Engineering Science Invention*, 2(11), 37–41.

Vinet, L. and Zhedanov, A. (2011). A "missing" family of classical orthogonal polynomials. *Journal of Physics A: Mathematical and Theoretical*, 44(8).

Wang, T. K., Zhang, Q., Chong, H. Y. and Wang, X. (2017). Integrated supplier selection framework in a resilient construction supply chain: An approach via analytic hierarchy process (AHP) and grey relational analysis (GRA). *Sustainability (Switzerland)*, 9(2).

Yu, C. and Matta, A. (2016). A statistical framework of data-driven bottleneck identification in manufacturing systems. *International Journal of Production Research*, 54(21), 6317–6332.

Yu, W., Jacobs, M. A., Chavez, R. and Feng, M. (2019). Data-driven supply chain orientation and financial performance: The moderating effect of innovation-focused complementary assets. *British Journal of Management*, 30(2), 299–314.

Yurdakul, M. and Çoğun, C. (2003). Development of a multi-attribute selection procedure for non-traditional machining processes. *Proceedings of the Institution of Mechanical Engineers, Part B: Journal of Engineering Manufacture*, 217(7), 993–1009.

Zardari, N. H., Ahmed, K., Shirazi, S. M. and Yusop, Z. B. (2014). *Weighting methods and their effects on multi-criteria decision making model outcomes in water resources management*. SpringerBriefs in Water Science and Technology, Springer.

Zeba, G., Dabić, M., Čičak, M., Daim, T. and Yalcin, H. (2021). Technology mining: Artificial intelligence in manufacturing. *Technological Forecasting and Social Change*, 171.

Zhong, R. Y., Xu, X., Klotz, E. and Newman, S. T. (2017). Intelligent manufacturing in the context of industry 4.0: A review. *Engineering*, 3(5), 616–630.

6 Entrepreneurial Venture Funding and Growth in Industry 4.0 Era

Deergha Sharma and Minakshi Sehrawat

CONTENTS

6.1 Introduction...89
6.2 Funding Avenues in a Start-Up's Life Cycle 90
6.3 Start-Up Environment...91
6.4 Emergence of Debt Avenues 93
6.5 Equity-Based Innovations in Entrepreneurial Financing..............................96
6.6 Government Initiatives for Fueling the Growth of Entrepreneurs...............101
6.7 Conclusion 102
References...102

6.1 INTRODUCTION

Economic progression has a close relationship with entrepreneurial growth and finance in the context of Industry 4.0. Understood as the backbone of any business, financial investments are channelized into productive avenues to stimulate the process of economic prosperity (Schumpeter and Backhaus, 2003). In the presence of robust financial markets, innovation has a significant role to play. Owing to the execution of innovative processes and commercialization of novel technologies, funds can be easily allocated to companies with the greatest potential and high efficiency (Kerr et al., 2015). More, to fund new ideas and technologies well-functioning financial markets shall exist as a first thing, followed by its capacity to lend, risk-taking range and percentage of payback loans (Scherer, 2011).

In developing economies, mature companies easily raise capital through bank loans and funds intermediated by public equity and bond markets. Although, financial constraints are critical for budding entrepreneurs specifically in the early and growth stages of the life cycle of a company when the business model is unproven. The survival of technocrats depends on their access to entrepreneurial finance in their pioneer stages and consequently to growth capital to advance their businesses. Over the years, financing novel business ideas has taken a paradigm shift with the advent of advanced funding avenues. Significant changes in financial technologies have transformed the way capital intermediation is implemented. The developments are influencing the firms in all phases of their life cycle.

DOI: 10.1201/9781003256663-6

The current article is an effort to discuss the emerging financing avenues for technocrats and initiatives taken by the Government to promote the growth of Entrepreneurs across various economies. The article is divided into various sections. The first section is the introduction followed by section two which has elaborated the life cycle of new ventures and viable funding options. The third section has discussed the ecosystem of start-ups. The debt and equity-based sources of finance are explained in sections four and five. Section 6 has highlighted several initiatives taken by the Government for promoting the growth of entrepreneurs while the conclusion is presented in section seven.

6.2   FUNDING AVENUES IN A START-UP'S LIFE CYCLE

Companies have access to several funding resources during their entire life cycle. Primarily, the entrepreneur's resources serve as a personal loan from the entrepreneur while holding levered equity claims in the company (Robb and Robinson, 2014; Bakhtiari et al., 2020). Moreover, lending is provided by relatives and friends, at times facilitated by Government grants or receives funding through different crowdfunding platforms. Although, the funds are generally not sufficient to survive in the current vibrant business environment. In the early phase, the cash inflows are hardly generated. As per The Global Entrepreneurship Monitor Report (2017), most of the businesses discontinue in this phase owing to lack of capital specifically in developing economies.

To raise capital, entrepreneurs are required to identify alternative avenues of funding. Considering the debt option, debt on credit cards, term loans from microfinance institutions, loans from Government institutions, venture debt, and crowdlending are some of the options available to technocrats. While taking the equity side into account, Venture Capital is extensively utilized as money of invention, which is extended by corporate venture capitalists and independent venture capital firms. However, the concentration of venture capitalists is now shifting from the early to the growth and expansion phases of new start-ups. Growth capital is often provided to budding entrepreneurs to facilitate the expansion of firms. This is specifically true in developing economies where firms are attempting to raise funds to test their business propositions in the widely developing markets. The shifting of funding from the early to the growth phase is bridged by an emerging group of angel investors and the proliferation of internet-based internet equity crowdfunding. Moreover, accelerators apart from providing capital also provide mentorship and significant networking opportunities to the founders.

For entrepreneurial projects that succeeded in approaching the growth phase, various financing modes are accessible in the expansion and mature stages. Apart from retained profits, it has been witnessed that banks are willing to extend funds as a firm has gathered tangible assets and showcased a feasible business model. In the growth phase, the technocrats can attain funds from non-conventional lenders, for instance, private credit funds. Likewise, sovereign wealth funds have also shown potential for funding untested viable projects. Moreover, by acquiring minority stake holding growth equity funds can be used for providing capital cushion to budding entrepreneurs. Some financial markets have a greater number of private equity-backed firms than public listed companies. Figure 6.1 depicts the mode of financing in different phases of the life of a firm.

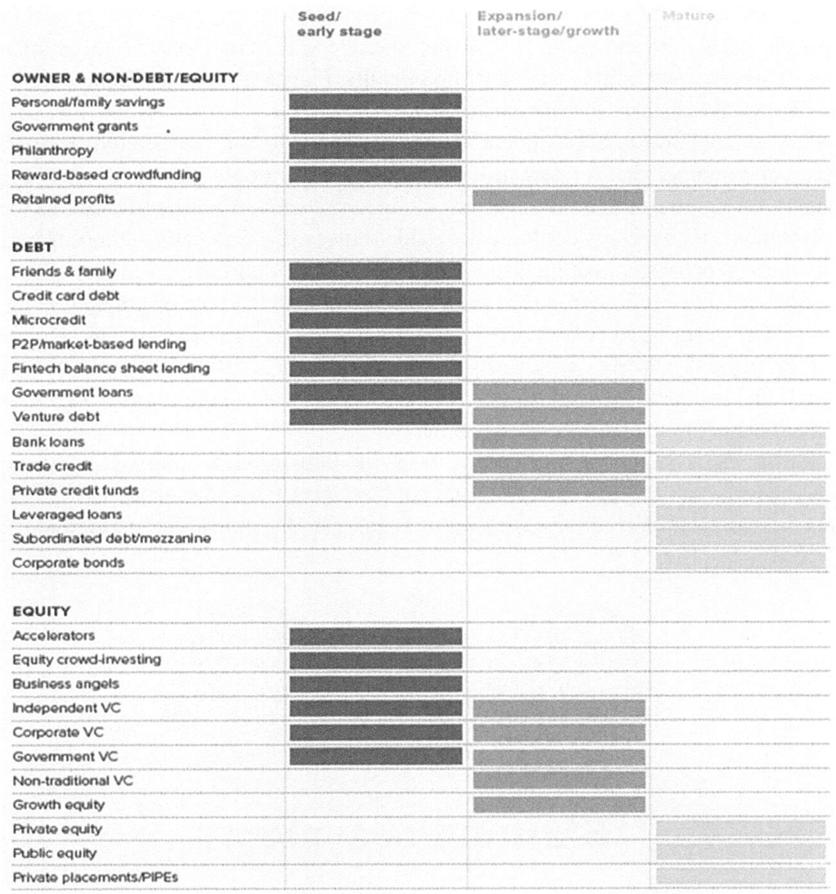

	Seed/ early stage	Expansion/ later-stage/growth	Mature
OWNER & NON-DEBT/EQUITY			
Personal/family savings	■■■		
Government grants	■■■		
Philanthropy	■■■		
Reward-based crowdfunding	■■■		
Retained profits		■■■	■
DEBT			
Friends & family	■■■		
Credit card debt	■■■		
Microcredit	■■■		
P2P/market-based lending	■■■		
Fintech balance sheet lending	■■■		
Government loans	■■■	■■■	
Venture debt	■■■	■■■	
Bank loans		■■■	■
Trade credit		■■■	■
Private credit funds		■■■	
Leveraged loans			■
Subordinated debt/mezzanine			■
Corporate bonds			■
EQUITY			
Accelerators	■■■		
Equity crowd-investing	■■■		
Business angels	■■■		
Independent VC	■■■	■■■	
Corporate VC	■■■	■■■	
Government VC	■■■	■■■	
Non-traditional VC		■■■	
Growth equity		■■■	
Private equity			■
Public equity			■
Private placements/PIPEs			■

FIGURE 6.1 Funding through avenues.

Source: Cornell University, INSEAD and WIPO, 2020

As companies attain maturity in their life cycle, the options of debt capital become more diverse particularly in developed economies in terms of corporate debt, subordinated debts, and mezzanine debt. The opportunity to go public facilitates access to retail and institutional investors. With the surge in investments in private equity funds, institutional investors have gained prominence in entrepreneurial finance.

6.3 START-UP ENVIRONMENT

The environment of start-ups is comprehended as manpower, numerous stages, and location, interacting in terms of a gadget to evolve new businesses. These corporations may be similarly categorized into universities, investment corporations, aid corporations (like incubators, accelerators, co-running areas, etc.), education, and support groups (like legal, monetary offerings, etc.), and big corporations. Different corporations normally pay attention to precise elements of the atmosphere characteristic and/

or start-ups at their precise improvement stage(s) spans across ideas, inventions and research, numerous stages, start-up group members, different enterprising humans from different corporations with start-up pursuits. Figure 6.2 has highlighted numerous factors of the start-up atmosphere.

People from those roles appear related collectively through shared activities, places, and interactions. As start-up ecosystems are commonly described with the aid of using the community of interactions among human beings, corporations, and their environment, many kinds could be available, however, are generally more referred to as start-up ecosystems of precise towns or online communities (even though a few can also say that because of social networks, the complete globe is simply one huge community of start-up ecosystems). In addition, assets like abilities, money, and time also are critical additives of a start-up environment. The assets that waft through ecosystems are received usually from human beings and corporations which might be lively a part of the start-up ecosystems.

Start-up ecosystems are managed with the aid of using each external and internal subsystems called components. External components such as financial environment, huge marketplace disruptions, and rapid organizations transitions manipulate the

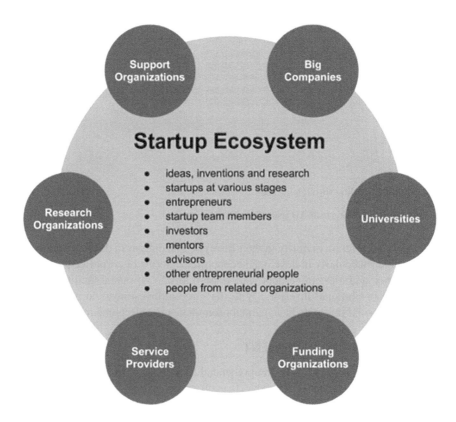

FIGURE 6.2   The Start-up ecosystem.

Source: Startup Commons, 2021

general shape of business environment. However, internal components provide the growth envision and create things to provide realistic outlook to the entrepreneurial journey. There can't be anything more valuable than exchanging inputs from external and internal components of this ecosystem. Naturally, it will help in reducing mistakes and making more nuanced business decisions.

While a number of useful input resources are commonly managed with the aid of using outside strategies like monetary weather and marketplace disruptions, the supply assets within the environment is managed with the aid of using inner elements like human beings and a corporation's cap potential to contribute closer to the environment. Although human beings exist and function inside ecosystems, their cumulative consequences are sufficient to steer outside elements like money market. Startup environments offer several items and offerings upon which different human beings and organizations rely and thus, the concepts of startup environment control advocate that as opposed to handling human beings or corporations, assets need to be controlled by the startup environment fundamentally.

Start-up environment control is pushed with the aid of using specific desires, achieved with the aid of using policies, protocols, and practices, and made adaptable with the aid of using tracking and studies primarily based on our excellent know-how of the interactions and strategies vital to maintaining environment systems and functions.

6.4   EMERGENCE OF DEBT AVENUES

The arising debt options for disbursing credit to the founders are as follows:

1) **Financial Technologies (FinTech)**: Fintech refers to a system that provides financial services on digital infrastructure. With fintech, the system is highly efficient, economical, transparent, and facilitates developing new ways of intermediation. The technology adopts artificial intelligence and machine learning, reduces the friction in the conventional lending market, and caters to the requirement of young technocrats. Owing to the larger role that technology plays in the financial sector, the FinTech lenders have broadened their horizons and offered endless possibilities. In the context of emerging markets, the usage of FinTech services has increased manifold. In Industry 4.0 the transformations in online banking, alternative lending platforms, and mobile banking have given impetus to the initiatives for tapping the unbanked population and provided substitutes to incumbent financial service providers. The widespread adoption of FinTech services has fueled consumerism, more efficient Government policies, and technological leapfrogging. The report of the Global Findex database has indicated that the proliferation of FinTech services is driven by digital payments, favorable regulatory framework, and innovative services assessed through mobile phones and advanced communication networks (Demirguc-Kunt et al., 2017). The adoption rate of financial technologies has been escalated to 64% globally while major growth drivers are countries like India and China with 87% usage followed by Russia and South Africa with 82% adoptions (EY FinTech Adoption Index, 2019). India

has Asia's highest FinTech investment activities (VC, PE, and M&A) with a deal value of around $647.5 Mn across ~33 deals, as compared to China's $284.9 Mn during the quarter ended June 30, 2020. The credit gap is filled by fintech lenders adopting business models depicted in Table 6.1.

According to the Global FinTech adoption Index 2019, 25% of SMEs have accepted FinTech services globally to address critical business challenges and provide relevant solutions. Figure 6.3 has shown some of the significant factors inducing the adoption of FinTech among the SME's operating worldwide. Further, Table 6.2 has shown emerging dimensions of FinTech to cater to the requirement of SMEs. Clyde, Digit, Flywire, Remitly, Chime, and Tala are some of the leading FinTech companies catering to the requirements of small-scale industries in emerging markets.

2) **Venture Debt**: It is the variation of debt financing for venture equity-backed companies lacking the assets for conventional avenues of debt financing. Being complementary to venture capital, it helps the entrepreneurs in building their businesses and achieving crucial milestones leading to better valuation during the fundraising process. The avenue employs non-convertible debenture as an underlying instrument which are coupon-bearing instruments issued by the debtor to the creditor. The facility of subscribing to the equity warrants is also available to the creditors. Owing to minimal equity dilution, the alternative fosters the growth of various start-ups. The financing is a boon for meeting working capital requirements and several

TABLE 6.1
Digital Lending Models

Digital Lending	Description
Transaction-based lending	Credit is extended using data of electronic transactions at POS.
Bank fintech partnership model	With bank tie-ups in specific segments such as travel, food, and hospitality, fintech companies source and underwrite potential borrowers for banks.
Invoice discounting exchanges	Operating exchanges where unpaid invoices can be discounted by SMEs to a network of financiers (Banks, NBFCs), wealth managers, and retail investors.
Marketplaces	Marketplaces like Paisa Bazaar connect borrowers with financial institutions. They provide the value add of digitizing the entire supply chain to provide borrowers with a seamless digital experience.
Captive models	Lending to a captive customer base either directly by setting up NBFCs (like Flipkart) or by partnering with financial institutions.
P2P model	Setting up P2P lending platforms to connect borrowers to affluent individuals with excessive liquidity.

Source: Administrator, 2019

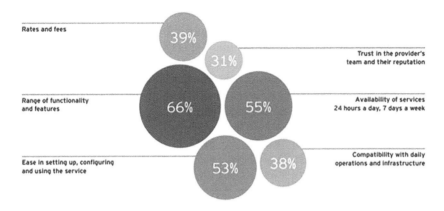

Rates and fees — 39%

31% — Trust in the provider's team and their reputation

Range of functionality and features — 66%

55% — Availability of services 24 hours a day, 7 days a week

Ease in setting up, configuring and using the service — 53%

38% — Compatibility with daily operations and infrastructure

FIGURE 6.3 Factors inducing FinTech adoption globally.

Source: Global FinTech adoption Index, 2019

TABLE 6.2
Emerging Dimensions of FinTech

Catergories	Services
Banking and payments	Online foreign exchange
	Digital-only branchless business bank
	Online payments processors
	Mobile Point of Sale (mPOS) payment machines and readers
Financial management	Online billing and invoice management tools
	Online cashflow and liquidity management tools
	Online bookkeeping and payroll tools
Financing	Online lending platforms
	Online marketplaces, aggregators and brokers
	Online equity (including equity crowdfunding) and debt securities
	Online invoice financing and dynamic discounting
Insurance	Insurance premium comparison sites

Source: Global FinTech adoption Index 2019

other capital expenditures when bank loans are not acting as a viable option for entrepreneurs. Accel, Sequoia Capital, Qualcomm Ventures, and Silicon Valley Bank are some of the leading venture capital debt firms extending the funds to several technocrats.

According to the report of Pitchbook (2021), the venture debt market has flourished, with venture capital-backed firms having taken more than $80 billion in loans and other debt options in the preceding three years. The venture (Stanford August 27, 2021) debt market has surged to a value of $28.2 billion in 2020. The major proportion of venture capital has been taken by tech companies who have taken around $18 billion in 2020. The report suggested that 11 over US 80 billion in loans and different debt merchandise have been originated for VC-subsidized organizations withinside the US given that 2018, with 2019 being a document year (US $28 billion) and 2020 now no longer away behind. In 2019, the undertaking enterprise deployed over the US $130 billion to US-primarily based organizations. Another instance is India, in which undertaking debt has been within the mainstream over the last 12 years. Venture debt investment is anticipated at about 8% of overall undertaking investment in 2020 (about US $800 million) developing from a base of among 4% to 5% of overall undertaking investment in preceding years. In Southeast Asia, VC investment within the vicinity has been developing progressively given that 2014, hitting a top in 2018 (US $12.6 billion) and appears set to have a document year in 2021 with Q1 already accomplishing US $6 billion well worth of fairness investment. Taking a 5% to 10% estimate of Southeast Asia's 3-year average (2018–2020) undertaking investment of US $9.8 billion in keeping with year, this shows that the capacity addressable undertaking debt marketplace in Southeast Asia can develop to among US $490 million to US $980 million annually. Venture debt is nicely positioned to develop in tandem with 2021 trying to be a document year for undertaking investment.

6.5   EQUITY-BASED INNOVATIONS IN ENTREPRENEURIAL FINANCING

Apart from debt options, some equity-based entrepreneurial funding options are as follows:

1) **Venture Capital**: It is an institutional or private investment in the form of equity, quasi-equity, participating debentures, and conditional loans in unlisted, high technology firms (Pandey,1998). The advent of Information and Communication Technology (ICT) has brought significant changes in ways credit is aiding budding entrepreneurs (Pradhan et al., 2019). The money is disbursed keeping in view the future potential to develop into highly profitable businesses. The venture capitalist provides management and business skills and becomes the financial partner of the enterprise by purchasing equity shares of the company. The cash is pooled from several investors and is utilized to fulfill the capital requirements of emerging companies with long-term growth. In the following stages, the funding is provided to the entrepreneurs:
    * **Seed Financing:** The stage represents the duration required by the technocrats to convert their idea into a business proposition. During this period, the entrepreneurs persuade their investors regarding the viability of their ideas. The funding covers the expenses of marketing research and product development intending to create a prototype for attracting further investments.
    * **Start-Up Stage:** This stage covers the funding required to fine-tune their proposed product and services, hire an essential workforce, and complete the research work if left.
    * **Emerging Stage:** The funding extended in the emerging stage is utilized by the start-ups for product manufacturing and to meet marketing and

sales expenses. The funds required in this stage are comparatively higher than in the seed and start-up stages.

- **Expansion Stage:** The financing provided in the expansion stage caters to the funding required for exponential growth and to meet current demands. This funding is also required for the expansion and diversification of business.
- **Bridge Stage:** The stage indicates the venture as a full-fledged viable business. The funding provided in this stage supports the activities like mergers and Initial Public Offerings. At this stage, the investors reap the benefits by selling the shares at a premium and obtaining a substantial return on their investments.

The funding through venture capitalists is increasingly enabling entrepreneurs to develop across borders in the era of Industry 4.0. Figure 6.4 shows the trend of venture capital financing worldwide. The trends have indicated the surge in deal value by quarter from 2013 until the second quarter of 2021. The growth of angel and seed financing, early and late-stage financing through venture capital is also displayed in the table endorsing the growth of development of venture capital funding. The funding has risen to $157.1 billion across 8,000 deals until the second quarter of 2021.

Table 6.3 is showing the global top 10 venture capital deals indicating the development of Venture Capital financing worldwide. It is evident from the table that Northvolt has attracted the highest $2.75 billion while Celonis, MessageBird, and Epic Games secured the funding of $1 billion through venture capital financing. Bessmer Venture Partners, Greycroft, Bain Capital Venture, Canaan Partners are some of the leading venture capital financing firms promoting technology know-how, development capabilities, and solvency of technocrats globally.

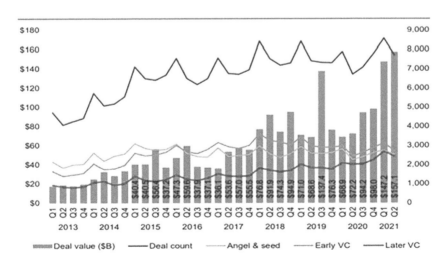

FIGURE 6.4 Global venture capital financing (2013–Q2'21).

Source: Venture Pulse Report—Global trends (2021)

TABLE 6.3

Venture Capital Proliferation Worldwide

S.No	Company	Deal Value (in billions)	Country	Sector	Type of Financing
1	Northvolt	$2.75	Sweden	Automotive cleantech	Series E
2	Waymo	$2.50	US	Automotive	Late-stage VC
3	J&T Express	$2	Indonesia	Logistics	Late-stage VC
4	BYJU'S	$1.55	India	Edtech	Late-stage VC
5	Nubank	$1.50	Brazil	FinTech	Series G
6	Horizon Robotics	$1.50	China	Semiconductors	Corporate
7	SpaceX	$1.20	US	Spacetech	Late-stage VC
8	Celonis	$1	Germany	Business/productivity software	Series D
9	MessageBird	$1	Netherlands	CloudTech	Series C
10	Epic Games	$1	US	Entertainment Software	Late-stage VC

Source: Q2'21 Venture Pulse Report- Global trends

2) **Angel Investing:** Angel investors act as the first source of external financing providing support to start-ups in the form of equity or debt in the early stages of business (Shane, 2008). This avenue is a boon for entrepreneurs when the probability of failure is significant, and investors are reluctant to extend funds to new businesses. Angel investors bridge the gap between small-scale financing usually contributed by family members and close relatives and venture capitalists (Sabarinathan, 2019). With an expectation of 25% to 60% return, angel investors are believed to approach more start-ups than venture capitalists (Bhide, 2003). Angel investors are usually entrepreneurs themselves and like venture capitalists, provide critical guidance and industry-specific insights to the entrepreneurs. The network of angel investors also paves the way for social entrepreneurship and establishes new business connections that facilitate the growth of a business (Ala-Jääski, and Puumalainen, 2021). Although, unlike venture capitalists, angel investors avoid investment in highly complex business proposals. Venture capital investments broadly fund high-tech start-ups while funding through angel investing has covered a broad range of industry sectors. Despite a risky investment, new entrants are seen as angel investors in India. Several high net-worth individuals and business owners of non-tech companies are funding the start-ups. Table 6.4 shows the key investments by angel investors globally through 2020.

3) **Incubation and Acceleration Programs:** Incubation and accelerator programs are one of the innovations in entrepreneurial finance providing short- and medium-term resources to start-ups and facilitating their progression

TABLE 6.4

Key Investments through Angel Investors and their Networks

S.No	Angel Investors	Key Investments
1	Ty Danco	Stitch, Localmind, Crashlytics, Codeship
2	Peter Kellner	Social Finance, LearnVest, Beepi, Contactually, Polymorph
3	Scott Banister	iLike, Zappos, Wavii, Vamo, Uber, ToutApp, Topsy Labs
4	Marh Cuban	The Mobile 360, Superfeedr, Soundwave, Ranku, Meta Saas
5	Wei Guo	YesGraph, Worklife, VetPronto, Verbling, TalkIQ
6	Paul Buchheit	lvl5, Zesty, Wufoo, Weebly, Virool, URX
7	Saad Alsogair	Abra, Onfleet, Ripple, Skykick Tovala, Wefunder, Spire
8	Talmadge O'NEILL	Abra, 500 Startups, Facebook, Hitch, LinkedIn, Mattermark
9	Naval Ravikant	Vurb, Visually, Uber, Twitter, Trusted, Topguest, Tinychat
10	Fabrice Grinda	Spotflux, eve Sleep, ecomom, Zesty, Xango.com, WiseStamp

of product development and time to market. Incubators and accelerators serve as the first contact for the entrepreneurs and help them in overcoming practical glitches. They organize various workshops and seminars and also host events such as talks by technical experts, informal networking meet-ups, and exposure visits to international ecosystems (Hausberg and Korreck, 2021).

Specifically, start-up incubators begin their journey with the technocrats who are in an early stage of business and not operating on a fixed schedule. Incubators are working independently or can be backed by venture capitalists, government entities, major corporations, or angel investors. The incubators help entrepreneurs refine their business proposals and build a foundation for their companies. In several cases, the start-ups under the incubation program are directed to a specific geographical area to operate with other companies under the incubation program. Within the incubator, the budding entrepreneurs refine their business ideas, clarify the product-market fit, analyze intellectual property challenges, and identify available networks in the ecosystems of start-ups. An incubator offers a co-working environment, additional mentoring, and a month-to-month lease program, and some networking with the local community. Highline beta, Start-up yard, Metavallon are some of the emerging incubators supporting the technocrats in their ventures.

While accelerators facilitate early-stage companies consisting of a promising Minimum Viable Product (MVP) with resources, education and mentorship, and connectivity with business partners and investors for rapid scale growth. Accelerators are fast-paced and intense preparing an early start-up market-ready in 3–6 months. In exchange for a small amount of equity, a seed investment is provided to the emerging companies along with access to a mentoring network comprising venture capitalists, industry experts, and start-up executives. Accelerators do not offer dedicated office space to the innovators but may provide a physical space for shared resources and organized events such as guest lectures. Y Combinator,

Techstars, and Brandy are some of the leading accelerators. Figure 6.5 shows the distinct features of accelerators and incubators. Further, Figure 6.6 depicts the rise of incubators and accelerators in some of the developing and developed economies with the inflection point in 2009.

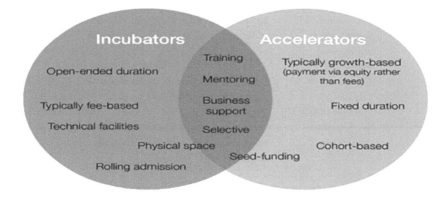

FIGURE 6.5   Features of accelerators and incubators.

Source: https://masschallenge.org/article/accelerators-vs-incubators

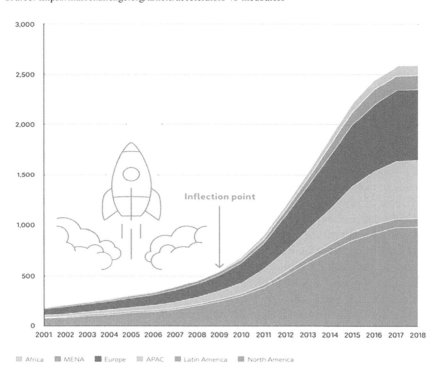

FIGURE 6.6   Growth of incubators and accelerators across developing and developed economies.

Source: Roland Berger

6.6 GOVERNMENT INITIATIVES FOR FUELING THE GROWTH OF ENTREPRENEURS

The steps taken by governments of various economies to enhance the viability of entrepreneurial ventures are discussed as follows:

- **United States of America:** The situation of shelter-in-place and the effect of unprecedented times has badly hampered small business units all over. According to the US Chamber, the US government has extended its helping hand to provide relief for millions of Americans through fiscal stimulus aid, government grants, government contract assistance, the Natural Resource Sales Assistance Program, and other small and sector-specific grants applicable to business depending on their size, time, and capacity. The popular Paycheck Protection Program is a loan designed to lessen the burden of small players with significant debt to pay off, provided they keep their employees on the payroll. It helps to deliver promising provisions of PPP Loans to small business owners with the help of its expert team of accountants and cash flow analysts. It is important to understand that the role of accountants in the whole process was foremost. They decide whether the business meets the required small business size and loan borrowing criteria or not. Accountants and analysts across companies own the responsibility of generating the cash flow and forecasting insights for their clients and help in determining the eligibility, and amount. Furthermore, after completely assisting the small business owners, it helped them by immediately doing the calculations on used and unused funds to leverage the affordable opportunity while keeping them protected in a time of need.
- **China:** Presently, with the Government incentives, China is the home of 4,300 creative spaces, around 3,300 incubators, and 400 accelerators growing exponentially every year. By inducing a healthy business environment for incubating and propagating start-ups, the country has approached the top two spots in terms of global venture capital (VC). Beijing, Shanghai, Shenzhen, Hangzhou, and Wuhan are some of the Chinese megacities nurturing several budding startups in the world. By relaxing tax regimes and liberal financial policies, the government is a pioneer in making China one of the best places to initiate a venture. Lower tax rates, tariffs, user friendly business registration procedure and focused approach for minority investors are some of the major attractions for entrepreneurs in China (Chiu, 2021).
- **Saudi Arabia:** The survey report of Global Entrepreneurship Monitor report (2019) indicated that around 76.3% of the adult population has found Saudi Arabia a favorable option with enormous opportunities to start a business. Notably, The Saudi Vision 2030 emphasized a thriving economy coupled with rewarding opportunities—aiming to stimulate the economy, amplify revenues which also accentuate SMEs as important agents of economic growth offering jobs, aiding innovation, and augmenting exports. The Saudi Vision 2030 is looking forward to raising the contribution of SMEs to Saudi Arabia's GDP from 20% to 35% by 2030. Recently, an incentive of US $19.2 billion was introduced by the government to uplift the private entities

emphasizing initiatives and schemes motivating SME sector for instance venture capital financing, indirect funding to budding entrepreneurs and Government fees reimbursement and capitalizing exports.

- **India:** According to IDFC's first bank report (2021), the Government has provided financing to startups and existing businesses considering capital to be the most critical factor for the existence of the enterprise. Besides supporting the entrepreneurs with angel funding, equity, and debt funding, IPOs, etc. the government also provides subsidies under various schemes for entrepreneurs to float their businesses. Further, the government of India has supported the informal entrepreneurs to start their ventures and grow them. The digital database made it convenient for firms to conduct business in an organized manner. As of 2021, there are an estimated 47 active investors in India, out of which venture capital firms constitute a 42% share. The first quarter of 2021 saw funding of $2.7 billion to Indian start-ups across 268 funding deals. Fintech was the most favored start-up and accounted for a 26% increase in funding deals.

6.7 CONCLUSION

In the present scenario of 4.0, the detailed plethora of alternatives available to foster the growth of entrepreneurs across the world are under researched. The present article has found answers around a variety of funding concerns, challenges, and wayouts in the context of emerging economies. An array of funding options available to entrepreneurs fosters their progress and sustains the several dynamic enterprises, in a credit-constrained scenario. The present article provides an overview of alternative avenue of entrepreneurial financing over the wide risk/return ambit and suggests that there exist opportunities to access increasingly composite and associated financial markets to cater the requirements of a highly assorted SME sector. The funding alternatives have supported the varied demand of SMEs in different phases. Although some are still in nascent stages of their life cycle or, in their present form, are approachable only to a small share of the SME community. In particular, investors prefer the businesses in mature stages while more funds are imperative to keep alive the start-ups in their early stages when the prototypes are required to be developed. It should be ensured that all industry sectors should be equally prioritized not just high-tech—encouraging growth across all industry sectors including low, mid, and high-tech firms. Moreover, active support to promising young start-ups by successful entrepreneurs will provide progress to new ventures. The favorable regulatory framework to facilitate easy lending to budding entrepreneurs would also stimulate unleashing growth of entrepreneurs. Overall, the new age of start-ups is looking for a more frictionless future with viable funding opportunities.

REFERENCES

Administrator. (2019). Fintech lending in India: Business models and future landscape. *TKWs Institute of Banking & Finance.* https://tkwsibf.edu.in/fintech-lending-in-india/
Ala-Jääski, S. and Puumalainen, K. (2021). Sharing a passion for the mission? Angel investing in social enterprises. *International Journal of Entrepreneurial Venturing*, 13(2), 165–185.

Bakhtiari, S., Breunig, R., Magnani, L. and Zhang, J. (2020). Financial constraints and small and medium enterprises: A review. *Economic Record*, 96(315), 506–523.

Bhide, A. (2003). *The origin and evolution of new businesses.* Oxford University Press.

Chiu, C. (2021, June 7). Why China is the best country to build your startup in 2022. *China Admissions.* www.china-admissions.com/blog/why-china-is-the-best-place-for-startups/

Cornell University, INSEAD and WIPO. (2020). *Global innovation index 2020: Who will finance innovation?* WIPO.

Demirguc-Kunt, A., Klapper, L., Singer, D. and Ansar, S. (2018). *The global Findex database 2017: Measuring financial inclusion and the Fintech revolution.* World Bank Publications.

EY FinTech Adoption Index. (2019). EY FinTech adoption index 2019, EY. https://www.ey.com/en_gl/ey-global-fintech-adoption-index (accessed date August 6, 2021)

GEM Global Entrepreneurship Monitor. (2017, April). GEM global entrepreneurship monitor. www.gemconsortium.org/report

Hausberg, J. P., & Korreck, S. (2021). Business incubators and accelerators: A co-citation analysis-based, systematic literature review. *Handbook of Research on Business and Technology Incubation and Acceleration*, Elgaronline, 39–63.

IDFC. (2021). Tracking entrepreneurship development in India. *IDFC FIRST Bank.* www.idfc firstbank.com/finfirst-blogs/beyond-banking/tracking-entrepreneurship-development-in-India

Kerr, S. P., Kerr, W. R. and Nanda, R. (2015). *House money and entrepreneurship* (No. w21458). National Bureau of Economic Research.

Pandey, I. (1998). The process of developing venture capital in India. *Technovation*, 18(4), 253–261.

PitchBook. (2021). Venture debt growth reaching all areas of VC market. *PitchBook.* https://pitchbook.com/newsletter/venture-debt-growth-reaching-all-areas-of-vc-market-bzd

Pradhan, R. P., Arvin, M. B., Nair, M., Bennett, S. E. and Bahmani, S. (2019). Short-term and long-term dynamics of venture capital and economic growth in a digital economy: A study of European countries. *Technology in Society*, 57, 125–134.

Robb, A. M. and Robinson, D. T. (2014). The capital structure decisions of new firms. *The Review of Financial Studies*, 27(1), 153–179.

Sabarinathan, G. (2019). Angel investments in India–trends, prospects and issues. *IIMB Management Review*, 31(2), 200–214.

Scherer, F. M. (2011). *New perspectives on economic growth and technological innovation.* Brookings Institution Press.

Schumpeter, J. and Backhaus, U. (2003). *The theory of economic development.* Joseph Alois Schumpeter, 61–116.

Shane, S. (2008). *Undefined.* Oxford University Press.

Startup Commons. (2021). What is startup ecosystem. *Startup Commons.* www.startupcommons.org/what-is-startup-ecosystem.html

7 Future Challenges and Opportunities in Adopting Industry 4.0 for Entrepreneurship

Rajeev Saha and Om Prakash Mishra

CONTENTS

7.1 Introduction 106
7.2 Industry 4.0 Technologies 107
    7.2.1 Cyber-Physical System (CPS) 107
    7.2.2 Internet of Things (IoT) 108
    7.2.3 Internet of Services (IoS) 109
    7.2.4 Big Data (BD) 109
    7.2.5 Cloud Computing (CC) 110
    7.2.6 Embedded System (ES) 110
    7.2.7 Augmented Reality (AR) 110
    7.2.8 Virtual Reality (VR) 110
    7.2.9 Information and Communication Technology (ICT) 110
    7.2.10 Artificial Intelligence (AI) 111
    7.2.11 Learning System (LS) 111
    7.2.12 Automated Manufacturing (AM) 111
    7.2.13 Cyber Security (CS) 111
    7.2.14 Collaborative Robots (CR) 113
    7.2.15 Flexible Manufacturing System (FMS) 113
7.3 Challenges in Adoption 113
    7.3.1 International Standards 113
    7.3.2 Integration of Various Technologies 113
    7.3.3 Interoperability 114
    7.3.4 Resistance to Change 114
    7.3.5 Initial High Cost 114
    7.3.6 Security Issues 114
    7.3.7 Big Data Handling 115
    7.3.8 Skilled Human Resource Availability 115

DOI: 10.1201/9781003256663-7

7.4 Opportunities for the Future 115
    7.4.1 Product Customization Possible in Less Time............................. 115
    7.4.2 Customer Driven Flexible Manufacturing 116
    7.4.3 Smart Monitoring of System 116
    7.4.4 Increased Return on Investment.. 116
    7.4.5 Better Supply Chain... 116
    7.4.6 Sustainable Socioeconomic Growth 116
7.5 Conclusion 117
References.. 119

7.1 INTRODUCTION

Industrialization has helped the world to move towards a better economic condition by producing entrepreneurs with new ideas and supporting people to live a better lifestyle. Entrepreneurs have to come up with new feasible ideas on a regular basis to sustain their industry. Advanced and integrative technologies of the Industry 4.0 ecosystem have given entrepreneurs a vision to create or upgrade their manufacturing facilities towards automation on a self-sustenance basis with the least direct human intervention.

Presently, most of the industries are following the industrial revolution named I-4.0. The first industrial revolution was reported in 18th century. At that time, steam engines were developed to generate power and this power was further utilized to develop the products. The industries were mechanized through the power generated. With extensive railroad network development, transportation of products and people became fast. The telegraph network helped in quickly transferring information, while electricity helped in developing modern factories. The invention of the assembly line was the next big thing helping industries to go for mass production. This was the time when the industries were striving to encash their resources in optimum manner. This era was named I-2.0 (Industry-2.0). The development of computers, program logic controllers, and allied items were reported in late 20th century. This era was named I-3.0 (Industry-3.0). During this era, the industries were shifting towards automation of the process to increase the production capacities (Singh et al., 2022). The latest revolution in the industrial sector is about the union of a physical system with advanced digital technologies called Industrial Revolution 4.0 (Dima, 2021). Figure 7.1 showcases the historical development of industrial revolutions. Germany is credited to initiate the concept of I-4.0. The concept was unveiled at the Hannover Messe in Germany and consists the cyber-physical systems wherein each constituent system will be integrated to form a network. (Kagermann et al., 2013). The constituents of the network will communicate with others in real-time thereby developing a self-reliant system that will interact and learn to self-adjust and adapt towards an optimized system.

With changing scenarios of the global business environment and rising competition, entrepreneurs are shifting from the traditional way of running industries to new business models specifically I-4.0. It was conceptualized for the transformation of existing industries into a digital factory and helped entrepreneurship grow to a transcendental level (Aiello et al., 2020; Oztemel and Gursev, 2020).

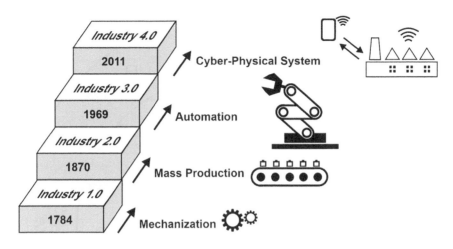

FIGURE 7.1 Historical perspective of Industrial Revolution

The technologies which are an integral part of Industry 4.0 needs to be inducted into constituent systems. The technologies prevalent world over include the Internet of Things for objects connectivity, Big Data Analytics for a predictive and prescriptive solution, Cloud for data capturing and handling, Machine Learning for adaptability of the machine towards better performance, Artificial Intelligence for making informed smart decisions, Augmented/Virtual Reality for the understanding complex working or structures, Robots for repetitive and tough jobs, Enterprise Resource Planning (ERP) for integrating back end working with the front end, 5G for communication, Additive Manufacturing for creation of a product in a short time, and Blockchain for secured information. Industry 4.0 is a new way of adding value to the business models by creating an automated and self-reliant system.

7.2 INDUSTRY 4.0 TECHNOLOGIES

Industry 4.0 came in existence due to collaborative technologies which enabled automation with almost no intervention of humans; the data to be captured in real-time from the entire value chain of a system and make informed decisions. The collaborative technologies (collection of core and allied technologies) are an integral embodiment of Industry 4.0 as shown in Figure 7.2. Industry 4.0 aims to automate the production system of a factory to a level where the operational part will be self-sufficient to optimize productivity and efficiency (Lu, 2017; Peruzzini et al., 2017).

7.2.1 CYBER-PHYSICAL SYSTEM (CPS)

An integrated system called a cyber-physical system (CPS) is created by combining the physical and digital worlds. It is provided with the ability to link and interact within and outside the system utilizing sensors, actuators, and internet technologies as needed (Dalenogare et al., 2018; Mittal et al., 2017; Vaidya et al., 2018;

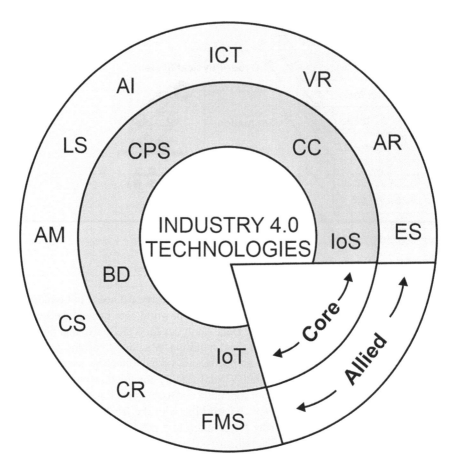

FIGURE 7.2   Collaborative technologies of Industry 4.0

Zhong et al., 2017). Figure 7.3 depicts a schematic representation of a cyber-physical system (Keil, 2017).

7.2.2 INTERNET OF THINGS (IOT)

Every object (thing) has its properties. The data relating to these properties gives us insight into various modalities of their working of the object (thing). The properties of the objects can be collected by attaching sensors. The connectivity of the object through the internet will help in collecting requisite data in real-time and analyzing them. The analysis and further feedback system in real-time will ultimately help in the improvement of the working of the concerned object. IoT integrates physical objects into information networks thereby converting them into smart objects (Da Costa et al., 2019; Rossit et al., 2018).

Within the IoT network, all things and humans are connected via the internet thereby being able to exchange information as per availability and requirement (Choi and Chung, 2017; Sadiku et al., 2017).

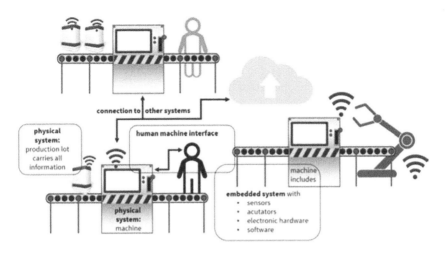

FIGURE 7.3   Structure of a CPS (Keil, 2017)

7.2.3   Internet of Services (IoS)

The streamlined functioning of any cyber-physical system is guaranteed only by the internet of services working in tandem with each other. The various services offered by IoS helps in running the factory through its entire value chain (Moreno-Vozmediano et al., 2012). The Internet of services (IoS) constitutes a service-oriented architecture and web 2.0 intermingled to reuse existing resources and services (Reis & Gonçalves, 2018). IoS helps interaction among connected things, humans, and systems. IoS provides internet-based services on demand for controlling the behavior of connected things (Anderl, 2014).

7.2.4   Big Data (BD)

The connected objects within the Industry 4.0 ecosystem generate lots of data that needs to be stored, processed, and analyzed for making smart decisions (Lu and Xu, 2019; Sbaglia et al., 2019; Wan et al., 2015). With the help of advanced computing technologies, large data sets of industries commonly called big data are analyzed helping industries to find trends and correlations for effective decision-making (Wang and Wang, 2016).

Big data is captured using sensors during manufacturing activities in smart factories. In Industry 4.0 scenario, the captured manufacturing data is being analyzed concurrently and used for the optimization of production as well as for predictive maintenance operation (Kim, 2017). Analysis of big data within Industry 4.0 helps in the optimization of operations management of an automated smart factory (Addo-Tenkorang and Help, 2016). Li et al. (2017) conducted a study on the use of analysis of big data collected from shop floor activities in a smart factory having Industry 4.0 practices. The study indicated improvement in overall efficiency of the smart factory due to analysis of big data and implementation of the results thereupon.

Within an Industry 4.0 setting, analysis of big data aids in the automation of production-related processes (Shanker et al., 2019; Castelo-Branco et al., 2019; Frank et al., 2019; Papadopoulos et al., 2022).

In an industrial environment, big data analytics can be combined with machine learning technologies to automate decision-making (Jamwal et al., 2022).

7.2.5   CLOUD COMPUTING (CC)

Cloud computing services include infrastructure, platform, and software as a service (Zhan et al., 2015; Zhong et al., 2017). Thus, cloud computing provides on-demand availability of computer system resources with functionalities distributed over multiple data centers.

7.2.6   EMBEDDED SYSTEM (ES)

Embedded systems are capable of monitoring and controlling physical and virtual processes within Industry 4.0 ecosystem through feedback cycles thereby enabling improvements in the system wherever required (Vaidya et al., 2018; Lee et al., 2015; Sindhwani et al., 2022).

An embedded system consisting of sensors, actuators, microprocessors/microcontrollers, and software helps in the collection of data from physical systems. Thus, an embedded system helps in coordinating physical things with computational elements in real-time within Industry 4.0 (Pereira and Romero, 2017).

7.2.7   AUGMENTED REALITY (AR)

The Augmented Reality (AR) is the advancement in the field of technology provides the real-time experience. The real-time experiences are developed by the machine itself based on the perception/knowledge based on visual or auditory and then converts the virtual environment in the real world. In the augmented reality, there is a mixture of both the virtual as well as real world in which the real-time interactions with physical objects is done and reported in three dimensions. This has the ability to be used as the modern-day technology in revealing the processes.

7.2.8   VIRTUAL REALITY (VR)

This is also the software-based technology that simulates the processes to enable the users to get accustomed to the real process (Mouf et al., 2018; Singh et al., 2022). The computer-generated environment in virtual reality can be explored and interacted to get a feel like real. The primary goal of VR is the ability to understand the real-world scenario of a system through its working replica of virtual world. The VR consistency and accuracy matters a lot because it dictates the virtual world.

7.2.9   INFORMATION AND COMMUNICATION TECHNOLOGY (ICT)

Each of the objects within the ecosystem of Industry 4.0 is generating some sort of data. This data or information needs to be communicated to the concerned object for its further utilization. ICT is useful in gathering information and communicating it through technologies developed for the purpose. ICT will help in automation process of an Industry 4.0 based system using real-time data exchange by including contemporary technologies, helping to build smarter end-to-end ecosystems (Peraković, Periša, & Zorić, 2019).

7.2.10 ARTIFICIAL INTELLIGENCE (AI)

Artificial Intelligence (AI) uses information, computer systems, and decision support systems to simulate the process of human intelligence. The industrial artificial intelligence are categorized into five different dimensions considering a. data; b. analytics; c. platform; d. operations, and e. human-machine technology. (Peres et al., 2020) Machine learning, natural language processing, and robotics are integral parts of industrial AI systems.

The AI-based Industry 4.0 operates autonomously in all areas collaborating with other autonomous systems within a defined boundary using a robotic system. The best part of the implementing AI in industries is, it helps in adapting the situations/scenarios communicated by the connected channels dynamically. AI is still evolving but has a tremendous future application within the context of Industry 4.0.

7.2.11 LEARNING SYSTEM (LS)

Learning systems utilizes the ongoing data accumulation and its analysis within a given manufacturing environment, thereby making smarter machines learn by themselves. It helps these machines to become self-reliant and decision-makers unto themselves. Machine Learning happens through a set of algorithms made for the purpose.

The data collected from the Industry 4.0 manufacturing environment besides helping in decision-making, is a continuous source of learning and knowledge enhancement as depicted in Figure 7.4 (Tvenge and Martinsen, 2018).

7.2.12 AUTOMATED MANUFACTURING (AM)

The entire manufacturing of products within Industry 4.0 will be automated. Automation helps in removing human-induced error in a manufacturing system. The productivity improves and hence the profitability. A smart factory is based on the concept of automated manufacturing. Figure 7.5 depicts the need for horizontal, vertical and end-to-end integration across the full value chain (Alcácer and Cruz-Machado, 2019).

7.2.13 CYBER SECURITY (CS)

With all objects interconnected and via the internet communicating with each other, the security issue is bound to crop up in Industry 4.0 ecosystem. A Cyber Security system is required to encrypt data and send it through secured paths. All care must be taken for ensuring the safety of the data-driven Industry 4.0 ecosystem.

ENISA, the European Union Agency for Network and Information Security, conducted a research to address the security and privacy issues associated with Industry

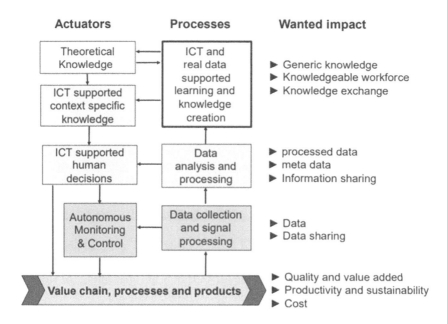

FIGURE 7.4   Learning system in Industry 4.0 (Tvenge and Martinsen, 2018).

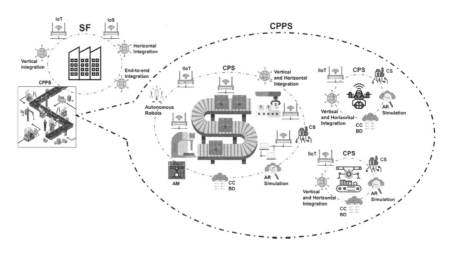

FIGURE 7.5   Smart Factory based on Industry 4.0 (Alcácer and Cruz-Machado, 2019)

4.0 collaborative technologies. ENISA conducted research on the security issues that Industry 4.0 devices and services face throughout their lifecycle. Based on this study, ENISA proposed security measures to be taken under three dimensions: policies, organizational measures, and technical measures (ENISA, 2018).

7.2.14 COLLABORATIVE ROBOTS (CR)

Collaborative Robots have the capability of interacting with each other along with the humans and machines within the manufacturing ecosystem (Vaidya et al., 2018). These Collaborative Robots are more formally called as Cobots.

Collaborative robots interact directly with human operators to execute tasks offering better flexibility and safety (Sherwani et al., 2020; Castillo et al., 2021). Standard ISO/TS 15066:2016 dealing with collaborative robots is already available. A new committee dedicated to robotics is working on ISO/TC 299 standards to frame standards for safety of industrial and service robots in light of new developments in the field of collaborative robots. Collaborative robots help in optimizing the environment. In addition, help the industrial setups in increasing the precision in the output through better quality.

7.2.15 FLEXIBLE MANUFACTURING SYSTEM (FMS)

With more and more customers demanding products as per their liking, it is quite predictable that in coming times the customer centric production system will be more popular. A flexible manufacturing system primarily takes customer demand into consideration. Flexibility in terms of time, operation and resources enables flexible and dynamic management of Industry 4.0 system benefitting its stakeholders (Gania et al., 2017).

7.3 CHALLENGES IN ADOPTION

There are several challenges explored below pertaining to the adoption of Industry 4.0 in present day entrepreneurial firms.

7.3.1 INTERNATIONAL STANDARDS

Though it's already a decade of introduction to Industry 4.0, there is no international standard in existence on Industry 4.0. For an industry to adopt the Industry 4.0 system, it needs to implement collaborating technologies working in tandem with each other. As each of the collaborating technologies has its own set standard and way of working, it becomes difficult to make them communicate and work with each other.

In Germany, DIN is involved in framing standards for Industry 4.0 and has taken great strides for making it easier for German companies to stay ahead (www.din.de, 2021). The international standards for Industry 4.0 will help different industries to adopt it with ease.

ISO Smart Manufacturing Coordinating Committee (2021) has proposed a three-part roadmap for standards in smart manufacturing. The three parts are enablers (≈enabling technologies), enhancers (≈design principles), and effects of smart manufacturing.

7.3.2 INTEGRATION OF VARIOUS TECHNOLOGIES

Core and allied technologies related to Industry 4.0 necessitates cross-industry collaboration. As each of these technologies were made for different purposes, their integration generally is being done on case-to-case basis as per requirement rather than as a standard system being followed internationally. Consider the core technologies CPS, IOT, IOS, BD, and CC. Integrating within an organizational system of these core technologies for desired output is a challenge. The earlier research reports that the vertical integration of information systems in the industries allowed the data flow in throughout the value chain. Still, the implementation of all technologies seems difficult to incorporate because of dependency on many attributes like decisions at various hierarchical levels, users with different capabilities, knowledge base, and cost of the system to be incorporated, and the industrial infrastructure etc. (Tabim et al., 2021).

7.3.3   INTEROPERABILITY

Systems within and across different industries needs to communicate with each other as per their requirement along the life cycle of a product (Liao et al., 2017). Technological parts of a system coming from different companies operates on different standards and protocols. As such their interoperability becomes a big opportunity. When such different systems are part of Industry 4.0 ecosystem, then they need to be framed as per a common standard for communicating with each other to interoperate.

7.3.4   RESISTANCE TO CHANGE

Humans are resistant to changes. Employee resistance to change is one of the most significant hurdles in implementing Industry 4.0 (Horvath and Szabo, 2019; Birkel et al., 2019). Ito (2021) recognized the specific sources of opposition to Industry 4.0 technologies being adopted at a rapid pace, as well as ways for dealing with them.

As Industry 4.0 entails huge changes with uncertainty of success during and after implementation, concerned people are bound to resist. Also, it is being seen more as a factor of increasing unemployment. Since in an Industry 4.0 scenario involvement of humans will be least, people across the globe take it against them. However, the case may be totally different. The number of jobs will shift more towards a white collared job rather than a blue collared job.

7.3.5   INITIAL HIGH COST

As the Industry 4.0 technology is still in the making, the initial cost is bound to be high, only to be risked by bigger companies. Some of the companies who have taken plunge into Industry 4.0 are IBM, Siemens, Schneider Electric, Airbus, SEAT, etc.

Unless initial cost comes down significantly, the medium and low valued enterprises may not dare to take the risk.

7.3.6   SECURITY ISSUES

Risk of compromising the entire system looms large when working within the Industry 4.0 ecosystem. As the system runs with the help of internet, the vulnerabilities in any of the equipment which is part of the system could make the system vulnerable to attack. The security threats thus are for real and finding a solution to the problem is not as easy.

Pereira et al. (2017) discuss about the new trend CYOD (Choose Your Own Device), which although introduces new technological changes, but at the same time increases security risks. Alani and Alloghani (2019) presents a detailed survey of threats looming within Industry 4.0 systems. Jamai (2020) discusses IIoT applications and related security issues in the Industry 4.0. Malatras et al. (2019) have documented high-level recommendations as per ENISA to promote Industry 4.0 cybersecurity among stakeholders in a secure manner.

It is quite evident that entrepreneurs must overcome security issue in their Industry 4.0 ecosystem to ensure uncompromised and uninterrupted manufacturing.

7.3.7 BIG DATA HANDLING

With vast amounts of data being produced by the equipment's and objects within Industry 4.0 ecosystem, it will be difficult to store such vast data for long duration. On the other hand, more data from different types and times will help in better analysis and machine leaning. Entrepreneurs will take help of data science and artificial intelligence to utilize this vast amount of data for optimization of processes and decision-making.

7.3.8 SKILLED HUMAN RESOURCE AVAILABILITY

As the focus in Industry 4.0 will be on automation, a skilled workforce is required to handle computing and automation related technologies. Vrchota et al. (2020) have classified necessary skillset required in Industry 4.0 on the basis of technical and personal abilities under three categories as "Must have," "Should have," and "Can have." "Must have" under "technical" trait includes IT skills, data analysis, statistics, awareness of organization and processes, and able to use latest devices. "Must have" under "personal" trait includes social and communication skills, work in team, self-management, time-management, and change adaptability. It will be a big challenge to train the workforce as per Industry 4.0 requirement since it requires a person to be trained with at least all of the traits in "Must have" as listed above.

7.4 OPPORTUNITIES FOR THE FUTURE

Despite several challenges, there are opportunities as well and explained below.

7.4.1 PRODUCT CUSTOMIZATION POSSIBLE IN LESS TIME

The product to be manufactured within Industry 4.0 environment can be customized as per market requirement in much less time. Consider the scenario without Industry 4.0 wherein a change in the product has been asked with quick replacement in the market. Unless a prototype is developed and tested properly with required changes, it may not be possible to quickly replace the existing product (Tyagi et al.,

2022). Additive manufacturing technology can be used to swiftly prepare a proto-type now that Industry 4.0 is in place. The prototype can then be tested virtually using simulation. FMS can further help in changing the manufacturing process as per requirement.

7.4.2 Customer Driven Flexible Manufacturing

With more options available with the customer today, they demand product as per their choice. Making products as per customer choice requires flexible manufacturing system which can quickly adjust the changes within its manufacturing ecosystem. Industry 4.0 has this capability of easily accommodating the manufacturing ecosystem as per requirements of the customer. With the delight of personalization choices, the trust and loyalty of the modern consumer increases.

7.4.3 Smart Monitoring of System

The components of Industry 4.0 based automated manufacturing system will monitor itself for any maintenance related activity. Instead of any breakdown, the components will be smart enough to take care of themselves wherever and whenever required (Sindhwani et al., 2021). It will help in continuing with the production system in the long run without worrying for any breakdown (Tyagi et al., 2022).

7.4.4 Increased Return on Investment

As a complete automated manufacturing system, the Industry 4.0 will have increased rate of production, thereby increasing profits and hence more return on investments. IoT and Cobots will ensure quality products being produced within a short span of time and delivered to the customers in quick succession thereby increasing return on investment. Figure 7.6 represents the economic impact on Europe on adoption of Industry 4.0 and its return on investment compared to a traditional industry (Think Act, Industry 4.0, 2014).

7.4.5 Better Supply Chain

With more real time data being made available, it will help in optimizing related partners of supply chain. Real time supply chain data will be available thereby ensuring the availability of the product at right place in right quantity. Delivery time to the customers will also decrease due to a better supply chain (Nagar et al., 2021).

7.4.6 Sustainable Socioeconomic Growth

Industry 4.0 introduces newer technologies to be adopted. The employees will be working with these newer technologies thereby increasing the overall productivity of the company. Newer technologies are expected to boost human capital demand by creating steady and well-paid positions in automated industrial areas, so contributing to the global fight against poverty (Evanthia, 2020).

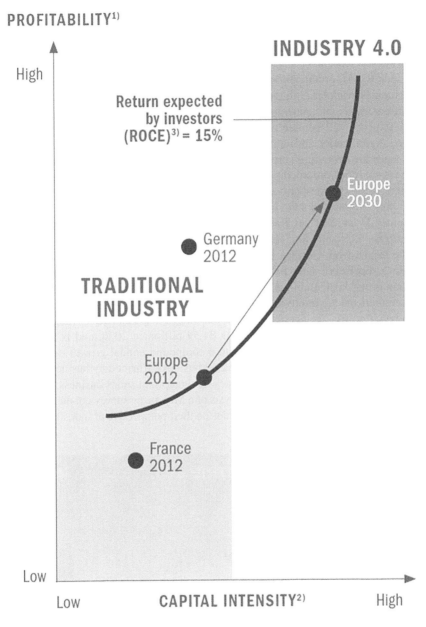

PROFITABILITY¹⁾

High

Return expected
by investors
(ROCE)³⁾ = 15%

INDUSTRY 4.0

Europe
2030

Germany
2012

**TRADITIONAL
INDUSTRY**

Europe
2012

France
2012

Low

Low   **CAPITAL INTENSITY²⁾**   High

FIGURE 7.6   Economic impact with Industry 4.0 (Think Act, Industry 4.0, 2014)

7.5   CONCLUSION

Industry 4.0 has provided the world a new opportunity of moving into a future of self-capable, smart, and intelligent production systems. However, the capabilities of Industry 4.0 cannot be harnessed completely unless various technologies work in harmony with each other and are standardized. The initial cost of setting up an Industry 4.0 based production system is too high and risky. With the product life cycle of new products becoming short and disruptive technologies making a foray into creating newer replacement products in a very short duration of time, it is quite risky to start with Industry 4.0 production system. Challenges are in abundance as of late, yet the opportunities available with Industry 4.0 seem to be more advantageous and realistic.

Entrepreneurs and start-ups taking the plunge into the transformation phase of manufacturing will be able to harness the advantages of the Industry 4.0 ecosystem. As digitalization and automation of manufacturing systems are required various companies are providing core technical solutions for cloud hosting, IIoT platforms, business analytics, microchips, sensors, hardware connectivity, cyber security, system integrators besides providing supporting technologies solutions for collaborative robots, augmented and virtual reality, automated transport vehicles, and drones. Siemens, GE, and PTC are leading the path providing industrial automation solutions for Industry 4.0 to budding entrepreneurs.

The results from German industries that adopted Industry 4.0 and reap the benefits in a short period set an example for others to follow. Figure 7.7 depicts the German firms' 2030 ambition for Industry 4.0.

Entrepreneurs are interested in technologies and business models which can guarantee better returns on investment. As reported by Facts and Factors (2020), Industry 4.0 had a global market value of USD 84.59 billion in 2020, and is predicted to reach USD 334.18 billion by 2028, with a compound annual growth rate of 19.4%. The growth potential of Industry 4.0 based business affirmed by business owners is attracting entrepreneurs to adopt Industry 4.0 compliant smart business.

Agrawal et al. (2021) showcase analysis of a McKinsey survey conducted on more than 400 global manufacturing companies on their perspective of Industry 4.0 during

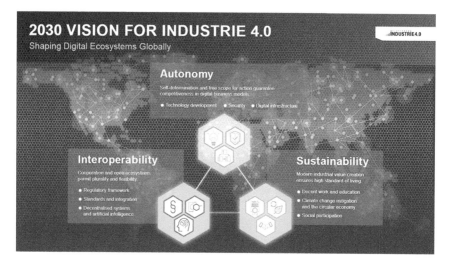

FIGURE 7.7   2030 vision for Industry 4.0 (www.plattform-i40.de/IP/Redaktion/EN/Infographics/vision2030.html)

the coronavirus pandemic. It was revealed that the adoption of Industry 4.0 technologies helped 94% of companies in operating their business during a pandemic. It was a wake-up call for others who have not implemented Industry 4.0 technologies and found themselves constrained in responding to the crisis.

A study conducted by PwC on global manufacturing companies adopting Industry 4.0, revealed that only 10% are fully compliant while 66% are in the transformation phase (PwC, 2019). As more and more business houses and entrepreneurs are realizing the fact that Industry 4.0 is the future to ensure being ahead and competitive, they are joining the fray.

REFERENCES

Addo-Tenkorang, R. and Helo, P. T. (2016). Big data applications in operations/supply-chain management: A literature review. *Computer and Industrial Enginering*, 101, 528–543.

Agrawal, M., Dutta, S., Kelly, R. and Millan, I. (2021). *COVID-19: An inflection point for Industry 4.0*. McKinsey & Company.

Aiello, G., Giallanza, A., Vacante, S., Fasoli, S. and Mascarella, G. (2020). Propulsion monitoring system for digitized ship management: Preliminary results from a case study. *Procedia Manufacturing*, 42, 16–23.

Alani M. M. and Alloghani M. (2019). Security challenges in the Industry 4.0 era. In *Industry 4.0 and engineering for a sustainable future*. Springer.

Alcácer, V. and Cruz-Machado, V. (2019). Scanning the Industry 4.0: A literature review on technologies for manufacturing systems. *Engineering Science and Technology, An International Journal*, 22, 899–919.

Anderl, R. (2014). Industrie 4.0 — Advanced Engineering of Smart Products and Smart Production, Technological Innovations in the Product Development 19th International Seminar on High Technology held at Piracicaba, Brasil.

Birkel, H. S., Veile, J. W., Müller, J. M., Hartmann, E. and Voigt, K. I. (2019). Development of a risk framework for Industry 4.0 in the context of sustainability for established manufacturers. *Sustainability*, 11(2), 384.

Castelo-Branco, I., Cruz-Jesus, F. and Oliveira, T. (2019). Assessing Industry 4.0 readiness in manufacturing: Evidence for the European Union. *Computer and Industrial Engineering*, 107, 22–32.

Castillo, J. F., Ortiz, J. H., Díaz, M. F. and Saavedra, D. F. (2021). COBOTS in Industry 4.0: Safe and efficient interaction. In *Collaborative and humanoid robots*. Intech-Open.

Choi, K. and Chung, S. H. (2017). Enhanced time-slotted channel hopping scheduling with quick setup time for industrial Internet of Things networks. *International Journal of Distributed Sensor Network*, 13(6).

Da Costa, M. B., Dos Santos, L. M. A. L., Schaefer, J. L., Baierle, I. C. and Nara, E. O. B. (2019). Industry 4.0 technologies basic network identification. *Scientometrics*, 121(2), 977–994.

Dalenogare, L. S., Benitez, G. B., Ayala, N. F. and Frank, A. G. (2018). The expected contribution of Industry 4.0 technologies for industrial performance. *International Journal of Production Economics*, 204, 383–394.

Dima, A. (2021). Short history of manufacturing: From Industry 1.0 to Industry 4.0. https://kfactory.eu/short-history-of-manufacturing-from-industry-1–0-to-industry-4–0/

European Union Agency for Network and Information Security (ENISA) (2018). *Good practices for security of internet of things in the context of smart manufacturing*. ENISA. ISBN: 978-92-9204-261-5, DOI: 10.2824/851384. https://www.enisa.europa.eu/publications/good-practices-for-security-of-iot

Evanthia, K. Z. (2020, January 21). *Fourth industrial revolution: Opportunities, challenges, and proposed policies, industrial robotics—new paradigms* (eds. A. Grau and Z. Wang). IntechOpen.

Facts and Factors (2020, March 27). Global Industry 4.0 market estimated to reach USD 334.18 billion by 2028. www.fnfresearch.com/news/global-industry-40-market-revenue-will-likely-exceed (accessed date February 5, 2022)

Frank, A. G. Dalenogare, L. S. and Ayala, N. F. (2019). Industry 4.0 technologies: Implementation patterns in manufacturing companies. *International Journal of Production Economies*, 210, 15–26.

Gania, I. P., Stachowiak, A. and Oleśków-Szłapka, J. (2017). Flexible manufacturing systems: Industry 4.0 solution. *Idea*, 10(1), 7–11.

Horvath, D. and Szabo, R. Z. (2019). Driving forces and barriers of industry 4.0: Do multinational and small and medium-sized companies have equal opportunities? *Technological Forecasting and Social Change*, 146, 119–132.

ISO Smart Manufacturing Coordinating Committee (2021, August 25). *White paper on smart manufacturing*. Charlotta Johnsson.

Ito, A., Ylipää, T., Gullander, P., Bokrantz, J., Centerholt, V. and Skoogh, A. (2021). Dealing with resistance to the use of Industry 4.0 technologies in production disturbance management. *Journal of Manufacturing Technology Management*, 32(9), 285–303.

Jamai, I., Azzouz, L. B. and Saïdane, L. A. (2020). Security issues in Industry 4.0. International Wireless Communications and Mobile Computing (IWCMC-2020), 481–488.

Jamwal, A., Agrawal, R., Sharma, M., Kumar, A., Kumar, V. and Garza-Reyes, J. A. A. (2022). Machine learning applications for sustainable manufacturing: A bibliometric-based review for future research. *Journal of Enterprise Information Management,* 35(2), 566–596.

Kagermann, H., Wahlster, H. and Helbig, J. (2013). *Securing the future of German manufacturing industry: Recommendations for implementing the strategic initiative INDUSTRIE 4.0—Final Report of the Industrie 4.0 working group.* Acatech—National Academy of Science and Engineering, 1–82.

Keil, S. (2017). Design of a cyber-physical production system for semiconductor manufacturing, In Digitalization in Supply Chain Management and Logistics: Smart and Digital Solutions for an Industry 4.0 Environment. Proceedings of the Hamburg International Conference of Logistics (HICL), GmbH, Berlin, 23, 319–340.

Kim, J. H. (2017). A review of cyber-physical system research relevant to the emerging IT trends: Industry 4.0, IoT, big data, and cloud computing. *Journal of Industry Integration Management*, 2, 175–191.

Lee, J., Bagheri, B. and Kao, H. A. (2015). A cyber-physical systems architecture for industry 4.0-based manufacturing systems. *Manufacturing Letters*, 3, 18–23.

Li, D., Tang, H., Wang, S. and Liu, C. (2017). A big data enabled load-balancing control for smart manufacturing of industry 4.0. *Cluster Computing*, 20, 1855–1864.

Liao, Y., Ramos, L. F. P., Saturno, M., Deschamps, F., Loures, E. F. R. and Szejka, A. L. (2017). The role of interoperability in the fourth industrial revolution era. *IFAC-PapersOnLine*, 50(1), 12434–12439,

Lu, Y. (2017). Industry 4.0: A survey on technologies, applications and open research issues. *Journal of Industry Information Integration*, 6, 1–10.

Lu, Y. and Xu, X. (2019). Cloud-based manufacturing equipment and big data analytics to enable on-demand manufacturing services. *Robotics and Computer-Integrated Manufacturing*, 57, 92–102.

Malatras, A., Skouloudi, C. and Koukounas, A. (2019). *Good practices for security of IoT in the context of smart manufacturing*. European Union Agency for Network and Information Security (ENISA).

Mittal, S., Khan, M. A., Romero, D. and Wuest, T. (2017). Smart manufacturing: characteristics, technologies and enabling factors. *Proceedings of the Institution of Mechanical Engineers, Part B: Journal of Engineering Manufacture*, 233(5), 1342–1361.

Moeuf, A., Pellerin, R., Lamouri, S., Tamayo-Giraldo, S. and Barbaray, R. (2018). The industrial management of SMEs in the era of industry 4.0. *International Journal of Production Research*, 56(3), 1118–1136.

Moreno-Vozmediano, R., Montero, R. S. and Llorente, I. M. (2012). Key challenges in cloud computing: Enabling the future internet of services. *IEEE Internet Computing*, 17(4), 18–25.

Nagar, D., Raghav, S., Bhardwaj, A. Kumar, R., Singh, P. L. and Sindhwani, R. (2021). Machine learning: Best way to sustain the supply chain in the era of industry 4.0. *Materials Today: Proceedings*, 47(13), 3676–3682.

Oztemel, E. and Gursev, S. (2020). Literature review of industry 4.0 and related technologies. *Journal of Intelligent Manufacturing*, 31, 127–182.

Papadopoulos, T., Singh, S. P., Spanaki, K., Gunasekaran, A. and Dubey, R. (2022). Towards the next generation of manufacturing: Implications of big data and digitalization in the context of industry 4.0. *Production Planning & Control*, 33(2–3), 101–104.

Peraković, D., Periša, M. and Zorić, P. (2019). Challenges and issues of ICT in industry 4.0. In *Design, simulation, manufacturing: The innovation exchange*. Springer, 259–269.

Pereira, A. C. and Romero, F. (2017). A review of the meanings and the implications of the industry 4.0 concept. *Procedia Manufacturing*, 13, 1206–1214.

Pereira, T., Barreto, L. and Amaral, A. (2017). Network and information security challenges within Industry 4.0 paradigm. *Procedia Manufacturing*, 13, 1253–1260.

Peres, R. S., Jia, X., Lee, J., Sun, K., Colombo, A. W. and Barata, J. (2020). Industrial artificial intelligence in industry 4.0-systematic review, challenges and outlook. *IEEE Access*, 8, 121–139.

Peruzzini, M., Grandi, F. and Pellicciari, M. (2017). Benchmarking of tools for user experience analysis in industry 4.0. *Procedia Manufacturing*, 11, 806–813.

PWC (2019). The lost workforce upskilling for the future. www.pwc.com/m1/en/world-government-summit/documents/wgs-lost-workforce.pdf

Reis, J. Z. and Gonçalves, R. F. (2018). The role of internet of services (IoS) on Industry 4.0 through the service oriented architecture (SOA). In IFIP International Conference on Advances in Production Management Systems (APMS), Seoul, South Korea, 20–26.

Rossit, D. A., Tohmé, F. and Frutos, M. (2018). Industry 4.0: Smart scheduling. *International Journal of Production Research*, 1–12.

Sadiku, M. N. O., Wang, Y., Cui, S. and Musa, S. M. (2017). Industrial internet of things. *International Journal of Advanced Scientific Research*, 3(11), 1–5.

Sbaglia, L., Giberti, H. and Silvestri, M. (2019). The cyber-physical systems within the industry 4.0 Framework. In *IFToMM ITALY 2018: Advances in Italian mechanism science*. Springer, 415–423.

Shanker, K., Shankar, R. and Sindhwani, R. (2019). Advances in industrial and production engineering. In *Select proceedings of FLAME 2018 book series*. Springer-Nature.

Sherwani, F., Asad, M. M. and Ibrahim, B. S. K. K. (2020). Collaborative robots and industrial revolution 4.0 (IR 4.0). International Conference on Emerging Trends in Smart Technologies (ICETST), 1–5.

Sindhwani, R., Kumar, R., Behl, A., Singh, P. L., Kumar, A. and Gupta, T. (2021). Modelling enablers of efficiency and sustainability of healthcare: A m-TISM approach. *Benchmarking: An International Journal*, 29(3), 767–792.

Singh, P. L., Sindhwani, R., Sharma, B. P., Srivastava, P., Rajpoot, P. and Kumar, R. (2022). Analyse the critical success factor of green manufacturing for achieving sustainability in automotive sector. In *Recent trends in industrial and production engineering*. Springer, 79–94.

Tabim, V. M., Ayala, N. F. and Frank, A. G. (2021). *Implementing vertical integration in the industry 4.0 journey: Which factors influence the process of information systems adoption?* Information System Front.

Think Act, Industry 4.0 (2014, March). *Pub.* Roland Berger Strategy Consultants GMBH.

Tvenge, N. and Martinsen, K. (2018). Integration of digital learning in industry 4.0. *Procedia Manufacturing*, 23, 261–266.

Tyagi, V., Kumar, R., Singh, P. L. and Shakkarwal, P. (2022). Barriers in designing and developing products by additive manufacturing for bio-mechanics systems in healthcare sector. *Materials Today: Proceedings*, 50(5), 1123–1128.

Vaidya, S., Ambad, P. and Bhosle, S. (2018). Industry 4.0—A glimpse. *Procedia Manufacturing*, 20(1), 233–238.

Vrchota, J., Mařiková, M., Řehoř, P., Rolínek, L. and Toušek, R. (2020). Human resources readiness for industry 4.0. *Journal of Open Innovation: Technology, Market, and Complexity*, 6(1), 3.

Wan, J., Cai, H. and Zhou, K. (2015). Industrie 4.0: Enabling technologies. *IEEE*, 135–140.

Wang, L. and Wang, G. (2016). Big data in cyber-physical systems, digital manufacturing and industry 4.0. *International Journal of Engineering and Manufacturing*, 6, 1–8.

Zhan, Z.-H., Liu, X. F., Gong, Y. J., Zhang, J., Chung, H. S. H. and Li, Y. (2015). Cloud computing resource scheduling and a survey of its evolutionary approaches. *ACM Computing Surveys (CSUR)*, 47(4), 63.

Zhong, R. Y., Xu, X., Klotz, E. and Newman, S. T. (2017). Intelligent manufacturing in the context of industry 4.0: A review. *Engineering*, 3(5), 616–630. www.din.de/resource/blob/65354/1bed7e8d800cd4712d7d1786584a7a3a/roadmap-i4-0-e-data.pdf

8 Case Study on Entrepreneurial Characteristics, Attitude, and Self-Employment Intention of Undergraduates in Industry 4.0 Context

*Pushkar Dubey,Parul Dubey
and Kailash Kumar Sahu*

CONTENTS

8.1 Introduction 124
8.2 Literature Review 124
    8.2.1 Entrepreneurial Characteristics 124
    8.2.2 Entrepreneurial Attitude 125
    8.2.3 Self-Employment Intention 125
    8.2.4 Incorporated Variables in the Study 125
8.3 Methodology 127
    8.3.1 Research Objectives 127
    8.3.2 Hypothesis 127
    8.3.3 Sampling and Data Collection 128
    8.3.4 Research Instrument 128
    8.3.5 Scale Validation 129
    8.3.6 Data Analysis 130
8.4 Analysis and Interpretation 130
    8.4.1 Entrepreneurial Attitude of Engineering Undergraduates 130
    8.4.2 Self-Employment Intention of Engineering Undergraduates 131
    8.4.3 Entrepreneurial Characteristics of Engineering Undergraduates 131
8.5 Discussion 133
    8.5.1 Contribution of the Study 134

DOI: 10.1201/9781003256663-8

8.6   Conclusion 135

Acknowledgement 135

References... 135

8.1   INTRODUCTION

The importance of startups in modern economies has grown significantly (Kuratko et al., 2015). Small and medium-sized enterprises (SMEs) are important drivers of innovation in Industry 4.0, and they also have a significant impact on the innovation activities of established companies (Rammer and Peters, 2015). However, there are numerous specifics about Industry 4.0 that could make the entrepreneurial process more difficult.

In nations like India, where the number of graduates hugely outnumbered the number of jobs, unemployment has become a major problem. According to the State of India's Environment (SoE) 2019 numbers, India's unemployment rate has climbed by approximately 1.9 times i.e., 7.6% in April 2019, which is the highest in the preceding two years (Pandey, 2019). A staggering 7.61% of people are unemployed in rural regions.

According to Bhagchandani (2017), talented and competent young people and graduates are becoming an ever-growing challenge, since they are always looking for work rather than beginning a new role. Self-financing is very fragmented, concentrated in a few locations, and extremely low in volume throughout the country, making it difficult for new and experienced entrepreneurs to get the assistance they need. Because India lacks a creative, inventive, and entrepreneurial culture, it can't become the most significant start-up ecosystem center.

Entrepreneurship and job creation in India have been elevated to the top of the national agenda. Engineering students should consider entrepreneurship as a career path. India's economic growth and development might be greatly aided by the use of science and technology (Chaudhary, 2018). In view of this, the present study seeks to identity the entrepreneurial characteristics, attitude, and self-employment intention among male and female engineering undergraduates enrolled at the several government and private technical colleges of Chhattisgarh state.

8.2   LITERATURE REVIEW

8.2.1   Entrepreneurial Characteristics

Entrepreneurial conduct has been linked to a variety of personality factors at various points in time. Entrepreneurs and non-entrepreneurs vary in a variety of personality qualities. Researchers continue to focus on the study of individual features and personality characteristics, which have attracted prominence in the past and the present (Robinson et al., 1991; Ho and Koh, 1992; Koh, 1996; Bakotic and Kruzic, 2010). Entrepreneurs have a set of features or attributes that are unique to them (Patel, 2015; Miller, 2020; Weedmark, 2020; Cizmeci, 2021). A person is said to have a characteristic if it is present in their behavior. Individuals have various characteristics that contribute to the development of their personalities. To characterize and concentrate on an individual's personality/psychological variables and attributes, a trait approach to entrepreneurship is used (McClelland, 1961; Brockhaus, 1980). Individuals with entrepreneurial qualities or characteristics, such as a desire for complete control over their job, a capacity to handle uncertainty and obstacles, and a strong sense of devotion to their work, are all examples (Mitton, 1989; Miller, 2020; Cizmeci, 2021). Individuals based on being grouped into several psychological qualities, such as a high demand for accomplishment, higher level of self-confidence, are more likely to take risks and be open to new ideas (Ho and Koh, 1992; Davidsson, 1989; Patel, 2015; Miller, 2020). New business endeavors are attracted by certain personality attributes (Ismail et al., 2009; Vaishali, 2022). Attraction, in the form of an intention, is a key predictor of actions (Ajzen, 1991).

8.2.2 ENTREPRENEURIAL ATTITUDE

Many thinkers and academics have emphasized the significance of an attitude in entrepreneurship (Leffler, 2020; Vamvaka et al., 2020), which may be used to get a deeper understanding of the field (Olson and Bosserman, 1984). There are two key techniques to figuring out someone's attitude. Attitude is described as a multi-dimensional approach in the second and a unidimensional construct in the first. Affective response is the only expression of the unidimensional method (Fishbein and Ajzen, 1975). Individual responses are based on three sorts of reactions, according to the multidimensional component: emotion, cognition, and conation. When you combine these three components, you get what's known as the "attitude model" (Katz and Stotland, 1959; Rosenberg et al., 1960; Chaiken and Stangor, 1987).

8.2.3 SELF-EMPLOYMENT INTENTION

Individuals who are considering going into business for themselves are said to have a "self-employment intention" (Krueger and Carsrud, 1993; Luthje and Franke, 2003; Drennan et al., 2005; Souitaris et al., 2007). Starting a new company or becoming an entrepreneur may be made easier and more effective when one is in a state of mind conducive to entrepreneurship (Moriano, 2011). Entrepreneurial intent is influenced by a variety of circumstances, including those mentioned above. Many studies have shown that individuals with an entrepreneurial mindset are more likely to pursue their dreams of starting their own businesses (Robinson et al., 1991; Luthje and Franke, 2003; Phan et al., 2002; Law, 2020; Jha, 2021). There is a lot of evidence that entrepreneurial intentions are critical to understanding the entrepreneurial process (Krueger and Carsrud, 1993; Scott and Twomey, 1988; Scherer et al., 1990; Brenner et al., 1991; Kolvereid, 1996; Rasheed, H. S. and Rasheed, B. Y. (2003).

8.2.4 INCORPORATED VARIABLES IN THE STUDY

  (a) *Entrepreneurial Characteristics:* An individual's collection of characteristics and talents that distinguishes them as an entrepreneur include those that they were born with or those they have gained throughout the course of their life and professional career. Entrepreneurs that are successful in their endeavours have characteristics such as uncertainty tolerance and self-sufficiency. Aside from that, they have the capacity to take risks and the willingness to do so in a safe and measured setting.

(b) Ambiguity tolerance: Tolerance for ambiguity, according to Stanley Budner (1962), is "a predisposition to perceive unclear circumstances as sources of harm."

(c) Self-sufficiency: The strength of one's beliefs in his/her ability to effectively fulfill the duties and activities of an entrepreneur was characterized by Chen et al. (1998) as entrepreneurial self-efficacy/self-sufficiency.

(d) Locus of control: Having a strong internal locus of control means that the individual believes that the activities he or she does will have an influence on the result (Rasheed, H. S. and Rasheed, B. Y., 2003).

(e) Risk-taking propensity: Entrepreneurs have a predisposition for taking risks, which sets them apart from their peers (Ahmed, 1985; Shane, 1996; Miner et al., 1989).

(f) Planning and organizing ability: The ability to plan and organize activities which helps an entrepreneur to manage every work associated with the desired results including time, tools, and resources etc. (Beetroot, 2020).

(g) Social Networking: People are lured to social networking sites because they have a strong need to interact with other people. Society views it as a kind of social behavior as well as a variety of personality characteristics (Taormina and Lao, 2007).

(h) Self-employment intention: It has been characterized in a number of various ways, including the intention to start a new business (Zhao et al., 2005), the intention to own a company, and the ambition to be self-employed (Zhao et al., 2005; Douglas and Shepherd, 2000).

(i) Entrepreneurial attitude: Ajzen (1987) defined attitude as a learned predisposition of an individual to behave favorably or unfavorably towards an action; its formation can occur either through past or prior experiences or perceptions accumulated over an individual's life span (Kuehn, 2008). In simple words, entrepreneurial attitude as their opinion about the abilities, adaptability, and actions in the entrepreneurial process.

(j) Achievement/accomplishment: Achievement refers to the feeling to achieve the desired outcome in the entrepreneurial process. McClelland stated that accomplishment motivation is the cornerstone to successful entrepreneurial action (Chell, 2008).

(k) Innovation: Innovation is described as the development of new goods, processes, markets, or organizational structures. In the business world, innovation is defined as the process of observing and responding to company operations in new and different ways (Kirton, 1978; Drucker, 1985).

(l) Personal control: It is the individual's impression of control and influence over their company that is the focus of the study of perceived personal control. Positivity about starting a business stems from one's own self-control (Robinson et al., 1991).

*(m) Self-esteem:*To succeed in a new business endeavor, those who think they have the appropriate skill and ability set are more likely to put in the necessary effort (Douglas and Shepherd, 2000).

8.3 METHODOLOGY

8.3.1 RESEARCH OBJECTIVES

1. To identify the entrepreneurial attitude of male and female engineering undergraduates enrolled in different government/private technical institutions of Chhattisgarh state.
2. To identify the self-employment intention of male and female engineering undergraduates enrolled in different government/private technical institutions of Chhattisgarh state.
3. To identify the entrepreneurial characteristics of male and female engineering undergraduates enrolled in different government/private technical institutions of Chhattisgarh state.

8.3.2 HYPOTHESIS

H1. There is no significant difference among the mean score of male and female students across achievement construct of entrepreneurial attitude dimension.

H2. There is no significant difference among the mean score of male and female students across innovation construct of entrepreneurial attitude dimension.

H3. There is no significant difference among the mean score of male and female students across personal control construct of entrepreneurial attitude dimension.

H4. There is no significant difference among the mean score of male and female students across self-esteem construct of entrepreneurial attitude dimension.

H5. There is no significant difference among the mean score of male and female students across overall entrepreneurial attitude dimension.

H6. There is no significant difference among the mean score of male and female students across self-employment intention dimension.

H7. There is no significant difference among the mean score of male and female students across ambiguity tolerance construct of entrepreneurial characteristics dimension.

H8. There is no significant difference among the mean score of male and female students across self—sufficiency construct of entrepreneurial characteristics dimension.

H9. There is no significant difference among the mean score of male and female students across locus of control construct of entrepreneurial characteristics dimension.

H10. There is no significant difference among the mean score of male and female students across risk taking construct of entrepreneurial characteristics dimension.

H11. There is no significant difference among the mean score of male and
female students across planning and organizing ability construct of entrepreneurial characteristics dimension.

H12. There is no significant difference among the mean score of male and
female students across social networking construct of entrepreneurial characteristics dimension.

H13. There is no significant difference among the mean score of male and
female students across overall entrepreneurial characteristics dimension.

8.3.3 SAMPLING AND DATA COLLECTION

Correlational research design is applied in the present study. Technical students
enrolled in the third and fourth year in different government and private institution
were selected to participate in the present study. About 1,245 questionnaires were
sent to participants, out of which 1,000 questionnaires were found usable (approximately 80.32% response rate). The data collection process was completed during
November 2019 to March 2020 (see Table 8.1).

8.3.4 RESEARCH INSTRUMENT

Selection, development, and finalization of research instrument is the essential part
of collecting right form of primary data from respondents. Similarly, the authors
followed the scientific process of developing and validating the research instrument
for the present study. Firstly, the authors thoroughly studied previous literature and
extracted a total of 11 constructs related to the present study objectives either completely adapted or adapted with certain modifications. Following that, the created
constructs were given to four subject experts for their feedback on the content development and for the purpose of gaining some useful suggestions. Finally, after obtaining minor revisions to some questions by experts, the authors performed a pilot study
with a small sample size of 50 participants to ensure that the content validity was not
affected. No modifications were asked by the participants, and hence, a 47-item questionnaire was finalized for the collection of primary data. The final constructs and
items are entrepreneurial attitude dimensions (i.e., achievement (6 items), innovation
(6 items), personal control (5 items), self-esteem (6 items)) adapted from Ismail et al.
(2013), entrepreneurial characteristics dimensions (i.e., ambiguity tolerance (3 items)
adapted from Bezzina (2010), self-sufficiency (3 items) adapted from Bezzina (2010),
locus of control (3 items) adapted from Bezzina (2010), risk-taking propensity (3 items)
adapted from Bezzina (2010), planning and organizing ability (3 items) adapted from
Schmidt and Bohnenberger (2009), Rocha and Freitas (2014), and Moraes et al. (2018),

TABLE 8.1
Description of Primary Data

Gender		Locality		Institution		Total
Male	Female	Rural	Urban	Government	Private	
429	571	434	566	236	764	1000

social networking (3 items) adapted from Schmidt and Bohnenberger (2009), and self-employment intention (6 items) adapted from Ismail et al. (2013).

8.3.5 SCALE VALIDATION

The present study applied partial least square confirmatory factor analysis to examine the primary data's reliability and validity measures (see Tables 8.2 and 8.3).

TABLE 8.2
Measurement Results

Factors	Cronbach's Alpha	Rho A	CR	AVE
Achievement	0.857	0.887	0.861	0.52
Innovation	0.882	0.884	0.881	0.554
Personal Control	0.844	0.846	0.844	0.52
Self Esteem	0.716	0.733	0.715	0.503
Self-Employment Intention	0.812	0.827	0.817	0.531
Ambiguity	0.782	0.786	0.777	0.516
Self Sufficiency	0.765	0.774	0.77	0.504
Locus of Control	0.702	0.705	0.786	0.526
Risk-Taking	0.738	0.74	0.739	0.586
Planning	0.781	0.786	0.783	0.519
Social Networking	0.725	0.726	0.724	0.557
Entrepreneurial Attitude	0.944	0.951	0.946	0.539
Entrepreneurial Characteristics	0.893	0.897	0.894	0.521

TABLE 8.3
Discriminant Validity

				Discriminant Validity (Fornell-Larcker Criterion)						
	A	AT	I	LOC	PC	POA	RTP	SEI	SE	SS	SN
A	**0.988**										
AT	0.887	**0.824**									
I	0.846	0.821	**0.917**								
LOC	0.721	0.718	0.744	**0.973**							
PC	0.759	0.778	0.841	0.962	**0.925**						
POA	0.693	0.66	0.714	0.865	0.721	**0.97**					
RTP	0.659	0.65	0.709	0.785	0.702	0.804	**0.832**				
SEI	0.491	0.562	0.59	0.707	0.621	0.647	0.744	**0.877**			
SE	0.635	0.63	0.613	0.776	0.656	0.734	0.697	0.738	**0.742**		
SS	0.466	0.556	0.564	0.703	0.613	0.642	0.663	0.697	0.619	**0.833**	
SN	0.448	0.493	0.51	0.571	0.557	0.617	0.546	0.656	0.551	0.636	**0.597**

[A = Achievement; AT = Ambiguity tolerance; I = Innovation; LOC = Locus of control; PC = Personal control; POA = Planning and organizing ability; RTP = Risk taking propensity; SEI = Self-employment intention; SE = Self-esteem; SS = Self-sufficiency; SN = Social networking]

(a) **Reliability measures:** Cronbach's alpha is used to assess internal consistency, and the result must be larger than 0.7. (Nunnally, 1978). Cronbach's alpha was found to be more than 0.7 in the current investigation (see Table 8.2). The value of Rho A also helps in measuring the reliability whose value also must be greater 0.7, and the authors found the value of Rho A for all the constructs above 0.7 (see Table 8.2).

(b) **Convergent validity:** In order to determine whether the numerous elements on a scale are in agreement, convergent validity is crucial. The composite reliability (CR) value must be larger than 0.7 in order to be considered a reliable indicator of internal consistency (Bagozzi and Yi, 1988; Hair, 2010). Table 8.2 shows that all constructs have a composite reliability over 0.7. Convergent validity can be assessed by measuring the average extracted variance (AVE). The AVE must be greater than or equal to 0.7 in order to meet the standards. Table 8.2 shows the AVE values over 0.7 for each of the constructs.

(c) **Discriminant validity:** Constructs' independence is determined by their discriminant validity. To attain construct validity, the discriminant validity value must be higher than 0.5. Table 8.3 indicates the value of discriminant validity above 0.5 for all the 11 constructs. Hence, it concludes that the present study possesses a satisfactory measurement model.

8.3.6 DATA ANALYSIS

Statistical analysis of the primary data in this study was carried out using SPSS v25 (licensed version) and Smart PLS 3 (trial version).

8.4 ANALYSIS AND INTERPRETATION

8.4.1 ENTREPRENEURIAL ATTITUDE OF ENGINEERING UNDERGRADUATES

It is observed from the Table 8.4 that female (N= 571) and male (N= 429) undergraduates have a difference in the mean score for achievement construct of entrepreneurial attitude dimension. The mean score of males (M = 33.284) was higher than that of females (M = 31.345). The mean score of males' and females' difference is statistically significant, t (998) = 4.168, p < 0.01. Hence, it is concluded that the mean score for males and females differs statistically across achievement construct of entrepreneurial attitude dimension.

In Table 8.4, female (N= 571) and male (N= 429) undergraduates have a difference in the mean score for the innovation construct of entrepreneurial attitude dimension. The mean score of males (M = 34.214) was higher than that of females (M = 31.187). The mean score of males' and females' differences is statistically significant, t (998) = 6.517, p < 0.01. Hence, it is concluded that the mean score for males and females differs statistically across innovation construct of entrepreneurial attitude dimension.

In Table 8.4, female (N= 571) and male (N= 429) undergraduates have a difference in the mean score for the personal control construct of the entrepreneurial attitude dimension. The mean score of males (M = 27.801) was higher than that of females

TABLE 8.4
t-test of Entrepreneurial Attitude as per Gender of Engineering Undergraduates

Variable	Group	N	Mean	t	df	p-value
Achievement	Girls	571	31.345	4.168	998	p < 0.01
	Boys	429	33.284			
Innovation	Girls	571	31.187	6.517	998	p < 0.01
	Boys	429	34.214			
Personal Control	Girls	571	25.289	6.463	998	p < 0.01
	Boys	429	27.801			
Self-esteem	Girls	571	30.099	4.995	998	p < 0.01
	Boys	429	32.009			
Entrepreneurial Attitude	Girls	571	117.921	6.162	998	p < 0.01
	Boys	429	127.31			

(M = 25.289). The mean score of males' and females' differences is statistically significant, t (998) = 6.463, p < 0.01. Hence, it is concluded that the mean score for males and females differs statistically across the personal control construct of entrepreneurial attitude dimension.

In Table 8.4, female (N= 571) and male (N= 429) undergraduates have a difference in the mean score for self-esteem construct of entrepreneurial attitude dimension. The mean score of males (M = 32.009) was higher than that of females (M = 30.099). The mean score of males' and females' differences is statistically significant, t (998) = 4.995, p < 0.01. Hence, it is concluded that the mean score for males and females differs statistically across self-esteem construct of entrepreneurial attitude dimension.

In Table 8.4, female (N= 571) and male (N= 429) undergraduates have a difference in the mean score for the entrepreneurial attitude dimension. The mean score of males (M = 127.310) was higher than that of females (M = 117.921). The mean score of males' and females' differences is statistically significant, t (998) = 6.162, p < 0.01. Hence, it is concluded that the mean score for males and females differs statistically across the overall entrepreneurial attitude dimension.

8.4.2 SELF-EMPLOYMENT INTENTION OF ENGINEERING UNDERGRADUATES

It is observed from the Table 8.5 that females (N= 571) and males (N= 429) have a difference in the mean score for the self-employment intention dimension. The mean score of males (M = 31.067) was higher than that of females (M = 30.677). The mean score of males' and females' differences is statistically insignificant, t (998) = 0.874, p > 0.05. Hence, it is concluded that the mean score for males and females does not differ significantly across the self-employment intention dimension.

8.4.3 ENTREPRENEURIAL CHARACTERISTICS OF ENGINEERING UNDERGRADUATES

It is observed from the Table 8.6 that female (N = 571) and male (N = 429) undergraduates have a difference in the mean score for the ambiguity tolerance construct

TABLE 8.5

t-test of Self-Employment Intention as per Gender of Engineering Undergraduates

Variable	Group	N	Mean	t	df	p-value
Self-Employment Intention	Girls	571	30.677	0.874	998	p > 0.05
	Boys	429	31.067			

TABLE 8.6

t-test of Engineering Characteristics as per Gender of Engineering Undergraduates

Variable	Group	N	Mean	t	df	p-value
Ambiguity Tolerance	Girls	571	14.961	3.373	998	p < 0.01
	Boys	429	15.676			
Self Sufficiency	Girls	571	16.043	2.856	998	p < 0.05
	Boys	429	16.673			
Locus of Control	Girls	571	15.315	4.404	998	p < 0.01
	Boys	429	16.345			
Risk-Taking	Girls	571	14.833	1.332	998	p > 0.05
	Boys	429	15.153			
Planning & Organizing	Girls	571	15.569	2.886	998	p < 0.05
	Boys	429	16.233			
Social Networking	Girls	571	14.695	4.82	998	p < 0.01
	Boys	429	15.799			
Entrepreneurial Characteristics	Girls	571	91.418	4.221	998	p < 0.01
	Boys	429	95.881			

of entrepreneurial characteristics dimension. The mean score of males (M = 15.676) was higher than that of females (M = 14.961). The difference in males' and females' mean scores is statistically significant, t (998) = 3.373, p < 0.01. Hence, it is concluded that the mean score for males and females differs statistically across the ambiguity tolerance construct of entrepreneurial characteristics dimension.

In Table 8.6, female (N = 571) and male (N = 429) undergraduates have differences in the mean score for self-sufficiency construct of entrepreneurial characteristics dimension. The mean score of males (M = 16.673) was higher than that of females (M = 16.043). The mean score of males' and females' differences is statistically significant, t (998) = 2.856, p< 0.05. Hence, it is concluded that the mean score for males and females differs statistically across self—sufficiency construct of entrepreneurial characteristics dimension.

In Table 8.6, female (N= 571) and male (N= 429) undergraduates have a difference in the mean score for the locus of control construct of entrepreneurial characteristics dimension. The mean score of males (M = 16.345) was higher than that of females

(M = 15.315). The mean score of males' and females' difference is statistically significant, t (998) = 4.404, p < 0.01. Hence, it is concluded that the mean score for males and females differs statistically across locus of control construct of entrepreneurial characteristics dimension.

In Table 8.6, female (N= 571) and male (N= 429) undergraduates have a difference in the mean score for the risk-taking construct of entrepreneurial characteristics dimension. The mean score of males (M = 15.153) was higher than that of females (M = 14.833). The difference in males' and females' mean scores is statistically insignificant, t (998) = 1.332, p > 0.05. Hence, it is concluded that the mean score for males and females does not differ significantly across the risk-taking construct of entrepreneurial characteristics dimension.

In Table 8.6, female (N= 571) and male (N= 429) undergraduates have differences in the mean score for planning and organizing ability construct of entrepreneurial characteristics dimension. The mean score of males (M =16.233) was higher than that of females (M = 15.569). The mean score of males' and females' differences is statistically significant, t (998) =2.886, p <0.05. Hence, it is concluded that the mean score for males and females differs statistically across planning and organizing ability construct of entrepreneurial characteristics dimension.

In Table 8.6, female (N= 571) and male (N= 429) undergraduates have differences in the mean score for the social networking construct of entrepreneurial characteristics dimension. The mean score of males (M =15.799) was higher than that of females (M = 14.695). The mean score of males' and females' differences is statistically significant, t (998) = 4.820, p <0.01. Hence, it is concluded that the mean score for males and females differs statistically across social networking construct of entrepreneurial characteristics dimension.

In Table 8.6, female (N= 571) and male (N= 429) undergraduates have a difference in the mean score for the entrepreneurial characteristics dimension. The mean score of males (M = 95.881) was higher than that of females (M = 91.418). The mean score of males' and females' differences is statistically significant, t (998) = 4.221, p < 0.01. Hence, it is concluded that the mean score for males and females differs statistically across the overall entrepreneurial characteristics dimension.

8.5   DISCUSSION

The results of H1 (achievement), H2 (innovation), H3 (personal control), H4 (self-esteem), H5 (entrepreneurial attitude) explained that entrepreneurial attitude dimensions did not support the hypotheses which means there is a significant difference in the mean score among males and females engineering undergraduates. Thus, it can be derived that entrepreneurial attitude is satisfactory available among males and females undergraduates. It also suggests that male undergraduates possess higher entrepreneurial attitude than females (Wang and Wong, 2004; Minniti and Nardone, 2007; Gaetsewe, 2018; Brush et al., 2018). Risk-taking behavior, self-esteem, innovation, and achievement variables does not exist in female undergraduates' personalities due to the unfavorable environment for self-employment. The conductive environment could change this picture, and they might become more active in different entrepreneurial activities.

However, the results of H6 (self-employment intention) explained that self-employment intention support the hypothesis which means there is no significant difference in the mean score among male and female engineering undergraduates. Thus, it can be derived that self-employment intention found no difference among male and female engineering undergraduates. It indicates that despite of having positive entrepreneurial attitude, self-employment intention is still lacking among under-educated because they don't perceive positive encouragement from the government and any other support such as family or educational institutions.

Moreover, the results of H7 (ambiguity tolerance), H8 (self-efficacy), H9 (locus of control), H11 (planning and organizing ability), H12 (social networking), and H13 (entrepreneurial characteristics) explained that entrepreneurial characteristic dimensions did not support the hypothesis which means that there is a significant difference in the mean score among male and female engineering undergraduates. Thus, it can be derived that entrepreneurial characteristics is significantly present among male and female undergraduates. It indicates that male and female undergraduates sufficiently possess entrepreneurial characteristics and looking for positive encouragement and opportunities. These entrepreneurial characteristics need more nourishment for creating self-employment intention in order to exploit opportunities in the times of Industry 4.0.

8.5.1 Contribution of the Study

In the times of Industry 4.0, it becomes imperative to assess the potential of young entrepreneurs in the countries like India, so that they could be transformed into future entrepreneurs. The findings of this research have implications for management theory and practice. The present study discusses that whether entrepreneurial characteristics, attitude, and self-employment intention is present among engineering undergraduates. Indeed, engineering undergraduates are more suitable for starting up a new venture as they have more know-how to different technicalities of the processes. The outcome revealed that entrepreneurial characteristics and attitude was found statistically significant among male and female undergraduates, but has no significant difference for self-employment intention variables among engineering undergraduates. The results explain that engineering undergraduates have entrepreneurial characteristics and attitude for starting up new ventures, but have no inclination for self-employment intention.

The present study addresses the biggest issue of emerging economies i.e., unemployment, and this can only be effectively tackled with the entrepreneurship approach among students who have all the necessary technical know-how about different processes i.e., engineering students. India is facing a severe case of unemployment in the country, and is in a dire need of employment for youth. Entrepreneurship can be the only option in which people will become job creator, rather than job seeker. This culture of finding suitable jobs needs to changed, and opening up new venture has to be inculcated into the system so that students become more aware and more confident over their characteristics and attitude, which eventually lead to positive self-employment intention.

The present study will not only help government for policy making, but it will also be fruitful for educational institutes to provide a necessary environment for nurturing technical students' characteristics and attitude for creating positive self-employment intention. This study will also emphasize structural changes in the education system for positively transforming technical/other students into potential entrepreneurs.

8.6 CONCLUSION

Entrepreneurship is the need of the hour, especially when global unemployment is continuously rising. Moreover, lakhs of engineering students also keep increasing and it has also become a big matter of concern. The present study identifies that the engineering undergraduates possess entrepreneurial characteristics and attitude for starting up new business, but do not have the right inclination towards self-employment intention. Thus, the government and concerned authorities such as educational institutions need to remark these outcomes as they can be future entrepreneurs and potentially contribute to Industry 4.0 for the growth and development of the country economy, if they are adequately nurtured and provide the right guidance for entrepreneurial success.

ACKNOWLEDGEMENT

This study is a piece of a comprehensive research project financed by the Indian Council of Social Sciences Research (ICSSR), New Delhi, under the IMPRESS scheme (File no.: IMPRESS/P42/85/18–19/ICSSR dated 20.03.2019). The author sincerely acknowledges ICSSR, New Delhi, for providing resources for accomplishing the research work.

REFERENCES

Ahmed, S. (1985). Nach, risk-taking propensity, locus of control and entrepreneurship. *Personality and Individual Differences*, 6(6), 781–782.

Ajzen, I. (1987). Attitudes, Traits, and Actions: Dispositional Prediction of Behavior in Personality and Social Psychology. *Advances in Experimental Social Psychology*, 20, 1–63. https://doi.org/10.1016/S0065-2601(08)60411-6

Ajzen, I. (1991). The theory of planned behavior. *Organizational Behavior and Human Decision Processes*, 50(2), 179–211.

Bagozzi, R. P. and Yi, Y. (1988). On the evaluation of structural equation models. *Journal of the Academy of Marketing Science*, 16(1), 74–94.

Bakotic, D. and Kruzic, D. (2010). Students' perceptions and intentions towards entrepreneurship: the empirical findings from Croatia. *The Business Review*, 14(2), 209–215

Beetroot (2020). Planning and organising. *Beetroot*. www.beetroot.com/graduate-jobs/careers-advice/planning-and-organising/#:~:text=to%20find%20out!,What%20is%20planning%20and%20organising%3F,your%20own%20tasks%20and%20time

Bezzina, F. (2010). Characteristics of the Maltese entrepreneur. *International Journal of Arts and Sciences*, 3(7), 292–312.

Bhagchandani, R. (2017). How India can up its startup game by bolstering student entrepreneurship. *Yourstory*. https://yourstory.com/2017/01/india-startup-game-bolstering-student-entrepreneurship

Brenner, O. C., Pringle, C. D. and Greenhaus, J. H. (1991). Perceived fulfillment of organizational employment versus. *Journal of Small Business Management*, 29(3), 62.

Brockhaus, R. H. (1980). Risk taking propensity of entrepreneurs. *Academy of Management Journal*, 23(3), 509–520.

Brush, C., Edelman, L. F., Manolova, T. and Welter, F. (2018). A gendered look at entrepreneurship ecosystems. *Small Business Economics*, 53(2), 393–408.

Chaiken, S. and Stangor, C. (1987). Attitudes and attitude change. *Annual Review of Psychology*, 38(1), 575–630.

Chaudhary, A. (2018). Why it is important to create and foster entrepreneurial mindset in India's high school students. *Yourstory.* https://yourstory.com/2018/02/entrepreneurship-minds.et-high-school-students?utm _pageloadtype = scroll

Chell, E., Wicklander, D. E., Sturman, S. G. and Hoover, L. W. (2008). *The entrepreneurial personality: A social construction.* Routledge.

Chen, C. C., Greene, P. G. and Crick, A. (1998). Does entrepreneurial self-efficacy distinguish entrepreneurs from managers? *Journal of Business Venturing*, 13(4), 295–316.

Chye Koh, H. (1996). Testing hypotheses of entrepreneurial characteristics. *Journal of Managerial Psychology*, 11(3), 12–25.

Cizmeci, D. (2021). Characteristics of an entrepreneur: 10 essential qualities. *Daglar Cizmeci.* https://daglar-cizmeci.com/characteristics-of-an-entrepreneur/ (accessed date February 23, 2022)

Davidsson, P. (1989). Entrepreneurship—And after? A study of growth willingness in small firms. *Journal of Business Venturing*, 4(3), 211–226.

Douglas, E. J. and Shepherd, D. A. (2000). Entrepreneurship as a utility maximizing response. *Journal of Business Venturing*, 15(3), 231–251.

Drennan, J., Kennedy, J. and Renfrow, P. (2005). Impact of childhood experiences on the development of entrepreneurial intentions. *The International Journal of Entrepreneurship and Innovation*, 6(4), 231–238.

Drucker, P. F. (1985). Principles of successful innovation. *Research Management*, 28(5), 10–12.

Fishbein, M. and Ajzen, I. (1975). *Belief, attitude, intention, and behavior: An introduction to theory and research.* Addison-Wesley.

Gaetsewe, T. (2018). *Determinants of self-employment in botswana working papers no. 46. botswana institute for development policy analysis.* Available at: Working Papers—BIDPA.

Hair, J. F., Black, W. C., Babin, B. J. and Anderson, R. E. (2010). *Multivariate data analysis: International version.* Pearson.

Ho, T. S. and Koh, H. C. (1992). Differences in psychological characteristics between entrepreneurially inclined and non-entrepreneurially inclined accounting graduates in Singapore. *Entrepreneurship, Innovation and Change: An International Journal*, 1(2), 243–254.

Ismail, M., Khalid, S. A., Othman, M., Jusoff, H., Rahman, N. A., Kassim, K. M. and Zain, R. S. (2009). Entrepreneurial intention among Malaysian undergraduates. *International Journal of Business and Management*, 4(10), 54–60.

Ismail, N., Jaffar, N. and Hooi, T. S. (2013). Using EAO model to predict the self-employment intentions among the universities' undergraduates in Malaysia. *International Journal of Trade, Economics and Finance*, 282–287.

Jha, S. (2021). Entrepreneurial mindset: An essential life skill. *Entrepreneur India.* www.entrepreneur.com/article/370068

Katz, D. and Stotland, E. (1959). A preliminary statement to a theory of attitude structure and change. *Psychology: A Study of a Science*, 3, 423–475.

Kirton, M. (1978). Field dependence and adaption-innovation theories. *Perceptual and Motor Skills*, 47(3), 1239–1245.

Kolvereid, L. (1996). Prediction of employment status choice intentions. *Entrepreneurship Theory and Practice*, 21(1), 47–58.

Krueger, N. F. and Carsrud, A. L. (1993). Entrepreneurial intentions: Applying the theory of planned behaviour. *Entrepreneurship & Regional Development*, 5(4), 315–330.

Kuehn, K. W. (2008). Entrepreneurial intentions research: Implications for entrepreneurship education. *Journal of Entrepreneurship Education*, 11, 87.

Kuratko, D. F., Morris, M. H. and Schindehutte, M. (2015). Understanding the dynamics of entrepreneurship through framework approaches. *Small Business Economics*, 45(1), 1–13.

Law, T. (2020). Entrepreneurial mindset: 20 ways to think like an entrepreneur. *OBERLO*. www.oberlo.in/blog/entrepreneurial-mindset

Leffler, E. (2020). An entrepreneurial attitude: Implications for teachers' leadership skills? *Leadership and Policy in Schools*, 19(4), 640–654.

Luthje, C. and Franke, N. (2003). The 'making' of an entrepreneur: Testing a model of entrepreneurial intent among engineering students at MIT. *R and D Management*, 33(2), 135–147.

McClelland, D. C. (1961). *The achieving society*. D. Van Norstrand Co Inc.

Miller, K. (2020). 10 characteristics of successful entrepreneurs. *Harvard Business School Online*. https://online.hbs.edu/blog/post/characteristics-of-successful-entrepreneurs

Miner, J. B., Smith, N. R. and Bracker, J. S. (1989). Role of entrepreneurial task motivation in the growth of technologically innovative firms. *Journal of Applied Psychology*, 74(4), 554–560.

Minniti, M. and Nardone, C. (2007). Being in someone else's shoes: The role of gender in nascent entrepreneurship. *Small Business Economics*, 28(2–3), 223–238.

Mitton, D. G. (1989). The compleat entrepreneur. *Entrepreneurship Theory and Practice*, 13(3), 9–20.

Moraes, G. H., Iizuka, E. S. and Pedro, M. (2018). Effects of entrepreneurial characteristics and University environment on entrepreneurial intention. *Revista de Administração Contemporânea*, 22(2), 226–248.

Moriano, J. A., Gorgievski, M., Laguna, M., Stephan, U. and Zarafshani, K. (2011). A cross-cultural approach to understanding entrepreneurial intention. *Journal of Career Development*, 39(2), 162–185.

Nunnally, J. C. (1978). An Overview of Psychological Measurement. In: Wolman, B. B. (eds) *Clinical Diagnosis of Mental Disorders*. Springer, Boston, MA. https://doi.org/10.1007/978-1-4684-2490-4_4

Olson, P. D. and Bosserman, D. A. (1984). Attributes of the entrepreneurial type. *Business Horizons*, 27(3), 53–56.

Pandey, K. (2019). India's unemployment rate doubled in two years: SoE in figures. *Down to Earth*. www.downtoearth.org.in/news/economy/india-s-unemployment-rate-doubled-in-two-years-soe-in-figures-64953

Patel, S. (2015). The 12 characteristics of successful entrepreneurs. *Entrepreneur India*. www.entrepreneur.com/article/250564

Phan, P. H., Wong, P. K. and Wang, C. K. (2002). Antecedents to entrepreneurship among university students in Singapore: Beliefs, attitudes and background. *Journal of Enterprising Culture*, 10(02), 151–174.

Rammer, C. and Peters, B. (2015). Innovation als Erfolgsfaktor Der deutschen Industrie? Der Beitrag von produkt- und Prozessinnovationen Zu Beschäftigung und Exporten. *Vierteljahrshefte zur Wirtschaftsforschung*, 84(1), 13–35.

Rasheed, H. S. and Rasheed, B. Y. (2003). Developing entrepreneurial characteristics in minority youth: The effects of education and enterprise experience. In *Ethnic entrepreneurship: Structure and process*. Emerald Group Publishing Limited.

Robinson, P. B., Stimpson, D. V., Huefner, J. C. and Hunt, H. K. (1991). An attitude approach to the prediction of entrepreneurship. *Entrepreneurship Theory and Practice*, 15(4), 13–32.

Rocha, E. L. C., Freitas, A. A. F. (2014), Evaluation of entrepreneurship teaching among university students through the entrepreneurial profile. *Contemporary Administration Magazine*, 18(4), 465–486.

Rosenberg, M. J., Hovland, C. I., McGuire, W. J., Abelson, R. P. and Brehm, J. W. (1960). *Attitude organization and change: An analysis of consistency among attitude components (Yales studies in attitude and communication)*, Vol. III. Yale University Press.

Scherer, R. F., Brodzinski, J. D. and Wiebe, F. A. (1990). Entrepreneur career selection and gender: A socialization approach. *Journal of Small Business Management*, 28(2), 37.

Schmidt, S. and Bohnenberger, M. C. (2009). Entrepreneurial profile and organizational performance/Perfil empreendedor e desempenho organizacional. *RAC-Revista de Administracao Contemporanea*, 13(3), 450–468.

Scott, M. G. and Twomey, D. F. (1988). The long-term supply of entrepreneurs: students' career aspirations in relation to entrepreneurship. *Journal of Small Business Management*, 26(4), 5.

Shane, S. (1996). Explaining variation in rates of entrepreneurship in the United States: 1899–1988. *Journal of Management*, 22(5), 747–781.

Souitaris, V., Zerbinati, S. and Al-Laham, A. (2007). Do entrepreneurship programmes raise entrepreneurial intention of science and engineering students? The effect of learning, inspiration and resources. *Journal of Business Venturing*, 22(4), 566–591.

Stanley Budner, N. Y. (1962). Intolerance of ambiguity as a personality variable1. *Journal of Personality*, 30(1), 29–50.

Taormina, R. J. and Kin-Mei Lao, S. (2007). Measuring Chinese entrepreneurial motivation. *International Journal of Entrepreneurial Behavior & Research*, 13(4), 200–221.

Vaishali (2022). Characteristics of an entrepreneur: 5 things to know. *IIM SKILLS*. https://iimskills.com/5-characteristics-of-an-entrepreneurs-by-iim-skills/

Vamvaka, V., Stoforos, C. and Palaskas, T. (2020). Attitude toward entrepreneurship, perceived behavioral control, and entrepreneurial intention: Dimensionality, structural relationships, and gender differences. *Journal of Innovative Entrepreneur*, 9(5).

Wang, C. K. and Wong, P. (2004). Entrepreneurial interest of university students in Singapore. *Technovation*, 24(2), 163–172.

Weedmark, D. (2020). What are three characteristics of an entrepreneur? *Small Business Chron*. https://smallbusiness.chron.com/three-characteristics-entrepreneur-18772.html (accessed date February 23, 2022)

Zhao, H., Seibert, S. E. and Hills, G. E. (2005). The mediating role of self-efficacy in the development of entrepreneurial intentions. *Journal of Applied Psychology*, 90(6), 1265–1272.

9 Study on Start-Ups Functioning in Industry 4.0 Context

Srijna Verma, Prateek Gaur, Rupali Madan and Vijay Kumar

CONTENTS

9.1 Introduction 139
9.2 Methodology 141
9.3 Successful Start-Up Stories 141
    9.3.1 Comparison among Reviewed Case Studies of Various Start-Ups 144
9.4 Identify Suitable Start-Ups Based on the Factors Using Intuitionistic Fuzzy Sets 145
    9.4.1 Case Study 146
9.5 Conclusion 149
References 149

9.1 INTRODUCTION

A company in its initial stages of operation is referred to as a start-up, which could be founded by one or more entrepreneurs. The level of innovation and invention a start-up offers is enormous as compared to other businesses (Blank and Dorf, 2020). Start-ups are imperative for entrepreneurial and economic growth of any nation (Corl, 2019; Rezaeinejad and Chernikov, 2021). The nations with maximum needs possess ample opportunities for start-ups to establish their economic growth. Boosting entrepreneurship is the only way to enhance the economic growth of any nation as it caters to employment and wealth generation. It comes with high risk but can be very rewarding (Corrales-Estrada, 2019). A small idea can convert into a successful start-up which can change the future of developing countries. Many successful companies like Microsoft, Apple etc. began as a start-up and converted into biggest giants in the areas where they are competing (Simple Search).

It is not an easy task to generate ideas that can be turned into successful ventures (Yang et al., 2018). Here, leveraging the new age technology, building digital strategies, considering sustainable factors are some of the important considerations behind the success of unusual ideas (Pangaribuan et al., 2021). Also, in the Industry 4.0 context, social media has played an important role and helping in deal with uncertain and unprintable situations. During the global pandemic, the technological sector boosts the economy and provides an opportunity to go-with entrepreneurial journey. Even though the Covid-19 pandemic had the detrimental impact on the economy leading to the shutdown of various start-ups, still fewer ones are working very well (Kuckertz et al., 2020). The reason behind this is their unique selling proposition that helped them to survive in such tough times and witnessed substantial growth within a year or two. In the present study, the start-up F possesses sustainability and skill-based features whereas another start-up A has language translation capabilities and promoting digitalization (Cassar, 2004). Start-up D has only the scalability feature. The output of the study will be helpful for the future aspirants to consider these as the base for decision-making in initiating the entrepreneurial journey. In addition, it is pertinent for entrepreneurs who should careful analyze the business environment which would help them to detect the untapped areas and even the uncertainties present in the market (Singh et al., 2022).

For running successful start-ups, there are several factors such as unique ideas, perfect timing, adequate funds etc. (Johnson, 2006; Ponomarev, 2019). There has been a dramatic growth in start-ups owing to technological, economic, and political factors (Ferreira and Lisboa, 2019). As far as start-ups are concerned, several case studies are reported to discuss on the results of defining new policies, strategies, or initiative in order to survive and thrive. This can be answered through providing an explanation to the questions of "how" "what" "why" in terms of implementation of unique idea or policy or strategy (Vaznyte and Andries, 2018; Almeida, 2020). It can even be useful in providing a great deal of information to the budding entrepreneurs about the areas that possess ample opportunities and the creative and unique ideas that made such cases a success (Caliendo and Kritikos, 2019).

While looking at the present context, it seems that the future lies in the hands of entrepreneurs who will provide jobs to the people. Even, the changing technology day-by-day transforming everything at very fast rate and making a good decision becomes the critical issue (Cohan, 2018). Though, many decision support tools have been developed, which have made the decision-makers' life easy and helped them to make decisions under favorable or unfavourable circumstances (De et al., 2001). Many decision-making techniques help in overcoming real life situations, which may be deterministic as well as non-deterministic in nature that helps in the journey of entrepreneurship. The global pandemic during the spread of Covid-19 is the best example to explain the difference between deterministic vs non-deterministic phenomena (Shanker et al., 2019; Salamzadeh and Dana, 2020; Camino-Mogro and Armijos, 2021). As far as pandemic is concerned, few start-ups such as Razorpay, Unacademy, and Nykaa beat the pandemic and became unicorns this year. Many start-ups have also emerged during such crucial times. Even, the economic growth during the pandemic reveals the funding to the start-ups was on great height (Startup boom in US during Covid-19, 2021).

The journeys of reputed start-ups reported in the literature reveal the importance of decision-making as they are more relying on it (Ardianti and Inggrid, 2018). Decision-making has applications in business, finance, management, economics, engineering, medical science, etc. It is very crucial when one has to select the better alternative from a finite number of available alternatives. Generally decision-making is the art or in other words the common activity used in every sphere of human functionality (Jain and Khandelwal, 2020; Nagar et al., 2021). Also, referred as the cognitive process upon which many studies (across various disciplines) have already been done to reveal the decision-maker's great interest towards solving the problems instead of creating them. Still, the ambiguous or imprecise information available in real life situations causes the functionality of the start-up (Ardito et al., 2015; Shen et al., 2020).

In the Industry 4.0 era, most of the entrepreneurs are learning the entrepreneurial facts and facets from the case studies that provide the description about the existing ventures/start-ups (Vaznyte and Anries, 2018; Sindhwani et al., 2021). These studies enable them to think and react on the important facts and facets rather than working on non-required activities. The discussion included in those case studies provides the intensive information about the start-ups and also on the founding team (Bergen and While, 2000; Muller et al., 2018).

The present day start-ups are now using both the deterministic process as well as non-deterministic processes for making effective decisions (Apsarini Soegoto, 2021). In the deterministic process, decision-making can be performed by means of an optimization problem. Where-as in the non-deterministic problems, making decisions under uncertainty is a challenge (Robertson et al., 2003; Fairlie et al., 2019). Under such circumstances, decision-makers face a lot of confusion and ambiguity in evaluating the criteria of alternatives. Here, it becomes pertinent to address the issue on what comes out from the start-up and how it can be accomplished? To encapsulate a wide range of perspectives unlike a single perspective it is essential for entrepreneurs to acquire exhaustive and unbiased knowledge on decision-making based on several factors either linked directly or indirectly (Perry, 2000; Shahab et al., 2019). The present study will explore various scopes for the growth of entrepreneurship and will provide useful insights to the budding entrepreneurs.

9.2 METHODOLOGY

The present study aims to reveal the growth and expansion opportunities for the new start-ups based on functioning characteristics (Innocenti and Zampi, 2019). For this purpose, the case studies on seven start-ups were taken anonymously and reviewed. The summarized reports on all the start-ups are given in Section 9.3. In these case studies, the important considerations are given on the idea (scope and boundaries); selection of problem (based on the support to targeted audiences); the quantitative and qualitative data to obtain for further processing and ultimately the interoperation based on analyses (The case study approach, 2011). Further, it is imperative to analyze the individual cases with the intentions to find out the future direction in the start-ups initiation and growth perspective especially in Industry 4.0 context using generalized fuzzy set based distance measure.

9.3 SUCCESSFUL START-UP STORIES

In the present study, the start-ups working in different sectors are identified anonymously based on their performance during the pandemic. Another fact for choosing these start-ups is the easy and uncomplicated process of delivering services to their targeted customers. The stories are summarized as below:

- **Persefoni climate technology:** With increasing pressure on businesses by investors to disclose their carbon footprints and climate concerns becoming a vital issue, a thought crossed Kentaro Kawamori (former chief digital officer at US shale gas company) mind. It was founded in January 2020 in Arizona and focused on a set of tools to gather, compute, and report organizational carbon emissions. The start-up provides the complete carbon footprint management system that facilitates enterprises and investors to manage, control, and report on their carbon outflows. This start-up leverages AI to provide management with information and data related to carbon emissions from their operations and allow them to measure and forecast their carbon transactions. The technology at Persofeni allows enterprises to reduce the environmental impact of their operations by turning raw data collected across different departments into actionable insights. It automates sustainability reporting by connecting reporting companies, shareholders, and institutional investors in a unified environment. Despite having only launched in January 2020, this Arizonian start-up was recently awarded as best technological innovator and won in the Enterprise and Data category at 13th annual SXSW pitch competition in March 2021 held by the Southwest event. Persefoni has advanced its leadership position in the climate tech industry with the aim to content worldwide climate change.
- **PayEm:** B2B pay-tech innovation is not going to stop as more and more executives have realized that it is necessary to keep pace with the global payments and that technology like PayEm's is must. It is a fin-tech start-up founded in 2019 and launched in March 2020 the same month when the World Health Organization declared Covid-19 a global pandemic. It is an expense management platform that provides automation, transparency, and control to finance teams. It automates monotonous tasks giving finance teams time to focus on more important work. It helps employees to keep track of all the spending of the team in one place and sync data easily with accounting software. The idea behind this start-up comes from an Israeli medical company that was developing an AI tool to diagnose rare diseases. Then, it was realized to undergo the loop holes in the credit card world. PayEm competes in a B2B payment market estimated at $127 trillion. The platform is very simple. All employees get a PayEm card for expenses and its rule and limit is set by the employer. PayEm's platform records receipts on the go with the PayEm app and generates expense reports automatically. The start-up's technology automates reimbursements, procurement, and credit card work flows. It also offers cross-border capabilities. Finance teams can make monetary transactions in 130 currencies over 200 territories in just one click. PayEm platform is a "one stop shop" for global monetary transactions. This fin-tech start-up allows employees and different departments in the organization to make their own spending decisions while providing centralized control and visibility for the finance teams to manage the finance of the company. The platform has seen colossal growth

in such a short span of time because of its ability to solve a fundamental problem of expense management.

- **EQRx:** EQRx, a biotech start-up launched in January 2020 with an unusual goal of developing low priced medications. Prior to this start-up, the founder has experience in medicine developments. The start-up intends to create and deliver novel medicines that are economical for patients and sustainable for health care systems. For this, the start-up has focused on reengineering the process of inventing new drugs for patient delivery while offering market-based solutions for soaring cost of medicines. They aim to create affordable medicines for patients and society. Leveraging the latest advances in science and technology the start-up seeks to devise, develop, and deliver good quality, patent protected and cost-effective medicines.

  The idea behind this was the growing opportunity for a company committed to inventing and introducing new drugs at different and lower prices than other drug makers charged (Sindhwani et al., 2022b. This start-up came at a point when payers and lawmakers across the United States were disappointed over high prices of some drugs covered by government funded insurance programs as the same drug was available at a lesser price to health care systems across the globe (Sindhwani et al., 2022a). The start-up targets the center of drug development including nucleic acids, protein, and other microscopic bits that drugs bind so as to become activated within the body. Each drug they develop would come at several hundred million dollars but if they go after known targets and use artificial intelligence, the probability of success will be higher. EQRx seeks to reproduce its vision at scale with the objective to launch one new drug in five years and ten in a decade.

- **Unschool:** In today's perspective, for the humans an outcome-based learning and personal mentorship is very important for meeting the personal as well as industry demands. Unschool aims to re-innovate learning and teaching methods according to on-going industry requirement/demand, while maintaining a personalized environment for each student so that they can learn easily (Singh et al., 2022). It was founded in 2019 and provides a platform where students can sign up easily by performing some simple steps and immediately start the course of their choice. Once they finish 20% of the course, they are involved with a minor project that tests their knowledge. After completing 70% of the selected course another major project intervention takes place. Once a student completes 80% of the course, then they are presented with handpicked internship opportunities from remarkable MNCs, NGOs etc. The unschool learning management system allows the host (Coach) to conduct their lessons in different formats, allow students to engage with coaches and peers and provides internship opportunities to them. It helps in the completion of live projects by providing personal mentorship from industry experts. It enables students to get placed in the best companies due to its result-oriented learning approach. It has been featured in LinkedIn's top start-ups list in September 2020 and been ranked third in India in the same year.

- **Nextbillion.ai:** A Singapore based start-up founded in 2020 aims to deliver a wide range of AI-powered hyperlocal solutions starting with the area of mapping. The start-up has already come out with its first product which is known as

Nextbillion Maps. This product is an AI-First SaaS map that provides customers with a range of APIs like routing, navigation, direction, and distance matrix. The founding team of this start-up was part of technology leadership team at Grab, a leading ride hailing app in Southeast Asia. At the initial stage, it was difficult for them to create fully friendly maps that every customer can use and get a perfect ETA all the time. Due to this people do not feel frustrated and a positive vibe will be used. The start-up is planning to consume the raised funds to expand this in North America and accessing the global markets. Currently the company has around 15 customers spread over 20 countries and has assisted companies in mapping over 2.5 million miles of roads.

- **JODO:** In the Covid-19 pandemic, many people lost their jobs and some families were not able to eat food two times a day. At that time, the middle- and low-income households suffered because their income became half and the expenses doubled. The school, colleges, institutes, transport, hospitals, daily need item price rose drastically and people with middle and low income were not able to manage their expenses. At that time, the JODO start-up provides them relief regarding the fees of their children. JODO is an Indian start-up and founded in 2020. It is a fin-tech based start-up that aims to target middle income households and helps them pay their children's school fees in installments without charging any kind of interest component on the payment. The start-up faced challenges in developing the network, like how they introduced their network channel to the middle-income people and schools regarding their proposal. The introduction of employee stock option plans at the early-stage fin-tech start-up helped the start-up to overcome the challenges of funds sources. Further it helps in accelerating the revenue growth up to 10x in a period of less than two years.

- **KOO APP:** KOO APP is a Twitter-like micro-blogging platform for Indian language speakers, it was launched in 2020. It is a user-friendly app that provides a very adaptive platform where one can also share their audio and video. The idea behind this app was to target Indian users who were facing challenges due to language barrier on other social media platforms. KOO APP became "*Vocal For Local*" after its launch user's enjoying it and around 4.2 million users registered through May 2020. Even, during the banning process for Chinese Apps by the Indian government, it works like a sling-shot for KOO, as they report their monthly active users surpassed 25 million. The start-up developed their platform friendly to Indian users just like language options and other features too that help each and every Indian to start blogging on their platform.

9.3.1 Comparison among Reviewed Case Studies of Various Start-Ups

This paper described seven start-up cases that have made their mark in the entrepreneurial world within a year or two. These cases have shown that leveraging presents opportunities by adapting business models to meet the needs of consumers and businesses proves to be successful. The Table 9.1 provides the comparison among the start-ups and identifies the suitable factors as:

TABLE 9.1
Comparison of Reviewed Start-ups

Start-Ups	Available in Industry	Unique Selling Proposition
Persefoni	Leveraging AI to recycle waste, real time supply chain carbon footprint management	Provides sustainability solutions by connecting reporting companies, shareholders, and institutional investors in a unified environment
PayEm	Cloud-based travelling expense platform, accounts receivable, and accounts payable management through software	"One stop shop" for global monetary safe transactions
EQRx	Machine learning for new drug discovery, transfigure biology into digital biology problem	Economical new drug discovery, affordable medicines
Unschool	Learn new skills by using online platforms so that each user can get whatever skills and knowledge they want	New way to assess users by completing different stages so that users can remember every part of their particular program
Nextbillion.ai	Satellite imaginary, aerial photography based maps were present	Provides AI-first SaaS maps that help customers with a range of APIs like routing, navigation, direction, and distance matrix
JODO	Cover up the expenses of the user and apply some interest to earn some profit	Pay the fees of children of middle income households in easy installments without any interest
KOO	Major regional languages available	More than 15 Indian regional languages offered

Source: Compiled by author

9.4 IDENTIFY SUITABLE START-UPS BASED ON THE FACTORS USING INTUITIONISTIC FUZZY SETS

It is a well-known fact for all kind of businesses that success is mainly depend on the decision-making process. In businesses, it is the toughest task to review and sort-out the voluminous data for getting the desired information which is further used as base in the decision-making (Kumar and Vashist, 2013). In start-ups functioning, large number of imprecise, incomplete, and vague information is available to make the decision which may cause the entire functioning of the start-ups. To deal with such situation, the Fuzzy Set Theory was introduced in 1965. It is the classical set theory generalization which has the capability to represent incomplete information about a situation in hand (Zadeh 1965). Even, more than five decades have been passed, still the researchers are using fuzzy sets for decision-making and developing new generalizations that makes decision-making process easier (Mehlawat and Gupta, 2014). In addition, the introduction to intuitionistic fuzzy set (IFS) which is the generalization of fuzzy set theory and characterized by the degree of membership/non-membership functions respectively (Kumar et al., 2012). These sets help in describing the fuzzy character more comprehensively in some special situations and therefore, are a more useful tool to deal with fuzzy information (Atanassov, 1989; De et al., 2001). The comparison of IFSs is usually done by using a similarity or distance measure between their associated memberships and/or non-membership functions. The Hamming and Euclidean distances are used to measure the distance between two sets (Kacprzyk et al., 1992; Szmidt and Kacprzyk, 2000).

The concepts of generalized fuzzy theory with the intuitionistic fuzzy sets are used for various decision-making methods across the business domains (Sarika et al., 2018). In the present study also, intuitionistic fuzzy based normalized Euclidean distance measure has been used for finding better start-ups based on the identified factors.

Let $W = \{w_1, w_2, ..., w_n\}$ be the discrete universe of discourse, the intuitionistic fuzzy set(IFS) is defined as: $F = \{< w, \mu_F(w), v_F(w) >| w \in W\}$, where

$\mu_F : W \rightarrow [0, 1], v_F : W \rightarrow [0, 1]$ and $\pi_F(w) = 1 - \mu_F(w) - v_F(w)$ are the membership function, non-membership function, and intuitionistic index of set F.

For measuring the distance in IFS, $W = \{w_1, w_2, ..., w_n\}$, then intuitionistic normalized Euclidean distance between the two intuitionistic fuzzy sets F and G is:

$$\Omega[F,G] = \left[\frac{1}{n} \sum_{i=1}^{n} \left[\left(\mu_F(w_i) - \mu_G(w_i) \right)^2 + \left(v_F(w_i) - v_G(w_i) \right)^2 + \left(\pi_F(w_i) - \pi_G(w_i) \right)^2 \right] \right]^{\frac{1}{2}}$$

The lowest distance measure between the two points is the decision value. The hypothetical case study has been framed to test the consistency.

9.4.1 CASE STUDY

Let $A = \{a_1, a_2, ..., a_m\}$, $B = \{b_1, b_2, ..., b_n\}$ and $C = \{c_1, c_2, ..., c_a\}$ be the finite sets of start-ups, entrepreneurship factors and general features in the start-ups respectively.

The intuitionistic fuzzy relations (IFR), P and Q are defined as:

$$P = \{< (c,b), \mu_P(c,b), v_P(c,b) >| (c,b) \in C \times B\}$$

$$Q = \{< (b,a), \mu_Q(b,a), v_Q(b,a) >| (b,a) \in B \times A\}$$

where,

(i) $\mu_p(C,b)$ and $v_p(C,b)$ indicate the degree to which the entrepreneurship factor [b] resembles and does not resemble the features of start-up [c] respectively.

(ii) $\mu_Q(b,a)$ and $v_Q(b,a)$ indicate the degree to which the entrepreneurship factor [b] confirm and does not confirm the start-up [a] respectively.

The research methodology is given in Figure 9.1.

In the present study, the entrepreneurship factors considered were government policies, social and cultural policies, business competition, economics and business environment, support system, training and critical thinking etc. On the other hand, the general features in start-ups are sustainability, scalability, skill based, language translation capabilities, and promoting digitalization etc. In our study, the start-ups

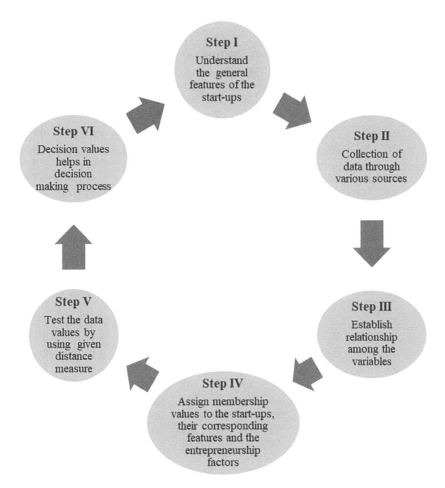

FIGURE 9.1 Flow chart of methodology.

considered as: Persefoni, PayEm, EQRx, Unschool, Nextbillion.ai, JODO, KOO. In the present context, the collection of useful data and organizing that one for information processing requires a considerable amount of time and money. Even, in social research maintaining the consistency and secrecy is also another challenge (Kumar and Vashist, 2013).

To test the consistency of the proposed model, hypothetical data has been taken in IF environment for demonstration purposes.

For this purpose, the IF values for general features of start-ups and their corresponding entrepreneurship factors are reported in Table 9.2. Further, the IF values for entrepreneurship factors and their likely start-ups is reported in Table 9.3.

From the information given in Table 9.2 and Table 9.3, the decision-making for the features present in start-ups c_i associated with the set of entrepreneurship factors b_j, characterize for each start-up has been presented by using intuitionistic normalized Euclidean distance measure. The lowest distance obtained points out correct decision-making as given in Table 9.4.

TABLE 9.2
IF Values for General Features of Start-ups and Their Corresponding Entrepreneurship Factors (IFR $P(C \rightarrow B)$)

P	b_1		b_2		b_3		b_4		b_5		b_6		b_7	
	μ_p	v_p	μ_p	v_p	μ_p	v_p	μ_p	v_p	μ_p	v_p	μ_p	v_p	μ_p	v_p
c_1	0.8	0.2	0.1	0.5	0.7	0.1	0.6	0.2	0.4	0.3	0.6	0.3	0.7	0.2
c_2	0.0	0.7	0.6	0.4	0.8	0.1	0.4	0.4	0.7	0.1	0.5	0.2	0.4	0.5
c_3	0.7	0.1	0.4	0.3	0.6	0.2	0.3	0.4	0.2	0.6	0.8	0.1	0.5	0.5
c_4	0.5	0.3	0.3	0.6	0.2	0.5	0.0	0.6	0.3	0.4	0.5	0.5	0.2	0.6
c_5	0.6	0.2	0.2	0.4	0.1	0.6	0.1	0.6	0.6	0.2	0.4	0.6	0.1	0.4

TABLE 9.3
IF Values for Entrepreneurship Factors and Their Likely Start-Ups (IFR $Q(B \rightarrow A)$)

Q	a_1		a_2		a_3		a_4		a_5		a_6		a_7	
	μ_Q	v_Q	μ_Q	v_Q	μ_Q	v_Q	μ_Q	v_Q	μ_Q	v_Q	μ_Q	v_Q	μ_Q	v_Q
b_1	0.4	0.4	0.3	0.6	0.2	0.7	0.1	0.7	0.4	0.4	0.5	0.4	0.2	0.6
b_2	0.3	0.6	0.1	0.8	0.4	0.5	0.6	0.2	0.2	0.3	0.1	0.7	0.6	0.2
b_3	0.2	0.7	0.6	0.3	0.2	0.6	0.5	0.3	0.7	0.2	0.6	0.2	0.4	0.5
b_4	0.1	0.8	0.2	0.7	0.7	0.2	0.4	0.4	0.3	0.6	0.4	0.5	0.7	0.1
b_5	0.4	0.6	0.6	0.3	0.3	0.5	0.5	0.4	0.5	0.4	0.2	0.7	0.8	0.1
b_6	0.5	0.2	0.4	0.5	0.1	0.7	0.3	0.6	0.4	0.3	0.7	0.1	0.1	0.7
b_7	0.2	0.6	0.5	0.5	0.6	0.2	0.2	0.7	0.1	0.6	0.2	0.6	0.7	0.1

TABLE 9.4
Distances Between the General Features Present in Start-Ups and the Related Start-Up

C_i	a_1	a_2	a_3	a_4	a_5	a_6	a_7
C_1	0.3257	0.1971	0.26	0.2942	0.1828	0.1571	0.2714
C_2	0.26	0.1428	0.2542	0.0999	0.1514	0.2256	0.1914
C_3	0.1914	0.2199	0.3257	0.2285	0.1428	0.0914	0.3828
C_4	0.0799	0.1285	0.2314	0.16	0.1542	0.1399	0.3442
C_5	0.1428	0.2028	0.2628	0.2157	0.1514	0.2428	0.2942

Based on the values in Table 9.4, the graph for minimum of the distance measure is developed and shown in Figure 9.2. The Figure 9.2 signifies the minimum distance measures for a particular feature which is useful for the start-up functioning.

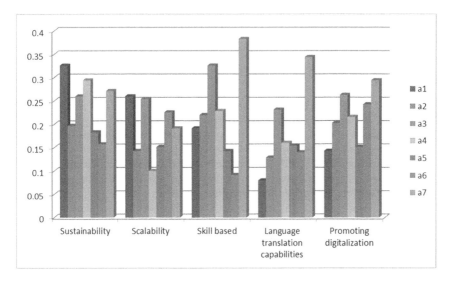

FIGURE 9.2   Comparison of distances between the general features present in Start-ups and the related Start-up.

9.5   RESULT AND DISCUSSION

Lesser value of distance measure, gives the estimation of the suitable alternative, discussed in the Figure 9.2. According to the proposed method, the start-up a6 has been shortlisted for the skill-based and sustainability features. For the features of Language translation capabilities and Promoting digitalization, start-up a1 has been shortlisted whereas, start-up a4 has been shortlisted for scalability. The proposed method has been used as a recommender tool for the recommendation of the suitable start-up for a particular task and has applications in the decision-making process.

9.6   CONCLUSION

In the competitive landscape for entrepreneurial strengthening, the start-ups regardless of size will undoubtedly develop entrepreneurial strategies all together contend to endure on the grounds. The published data on the journey of start-ups is taken as the input data for the present study. The data further described the basic strategies and opportunities, the start-ups considered to survive and sustain. The opportunities encashed by theses start-ups can be helpful when applied to similar kind of strut-ups in future. The main objective of this work is to establish the future directions for start-ups. For this reason, the study includes the cases of start-ups that flourished amid the pandemic. Still there are limitations that only a few entrepreneurial factors are considered. Even in real world scenario, there might be more factors available. A further study can be done to provide insights regarding the opportunities and challenges faced by the tech start-ups or technical sector. It will help budding entrepreneurs to know more about the trends in the tech start-up ecosystem. Also, it provides handheld support to decision-makers to promote entrepreneur activities in some segment by selecting the suitable start-up for a particular segment.

REFERENCES

Almeida, F. (2020). The role of tech startups in the fight against COVID-19. *World Journal of Science, Technology and Sustainable Development*, 18(1), 64–75.

Apsarini Soegoto, F. (2021). Product development using SWOT analysis. *International Journal of Entrepreneurship & Technopreneur*, 1(1), 1–10.

Ardianti, R. and Inggrid, N. (2018). Entrepreneurial motivation and entrepreneurial leadership of entrepreneurs: Evidence from the formal and informal economies. *International Journal of Entrepreneurship and Small Business*, 33(2), 159.

Ardito, L., Messeni Petruzelli, A. and Albino, V. (2015). From technological inventions to new products: A systematic review and research agenda of the main enabling factors. *European Management Review*, 12(3), 113–147.

Atanassov, K. T. (1989). More on intuitionistic fuzzy sets. *Fuzzy Sets and Systems*, 33(1), 37–45.

Bergen, A. and While, A. (2000). A case for case studies: Exploring the use of case study design in community nursing research. *Journal of Advanced Nursing*, 31(4), 926–934.

Blank, S. and Dorf, B. (2020). *The startup owner's manual: The step-by-step guide for building a great company.* John Wiley & Sons.

Caliendo, M. and Kritikos, A. S. (2019). "I want to, but I also need to": start-ups resulting from opportunity and necessity. In *From industrial organization to entrepreneurship*. Springer, 247–265.

Camino-Mogro, S. and Armijos, M. (2021). Short-term effects of COVID-19 lockdown on foreign direct investment: Evidence from ecuadorian firms. *Journal of International Development*, 34, 715–736.

Cassar, G. (2004). The financing of business start-UPS. *Journal of Business Venturing*, 19(2), 261–283.

Cohan, P. S. (2018). Boosting your startup common. In *Startup cities*. Apress, 219–235.

Corl, E. (2019). How startups drive the economy. *Medium*. https://medium.com/@ericcorl/how-startups-drive-the-economy-69b73cfbae1

Corrales-Estrada, M. (2019). *Innovation and entrepreneurship: A new mindset for emerging markets.* Emerald Publishing Limited, 267–278.

De, S. K., Biswas, R. and Roy, A. R. (2001). An application of Intuitionistic fuzzy sets in medical diagnosis. *Fuzzy Sets and Systems*, 117(2), 209–213.

Fairlie, R. W., Miranda, J. and Zolas, N. (2019). Measuring job creation, growth, and survival among the universe of Start-UPS in the United States using a combined start-up panel data set. *ILR Review*, 72(5), 1262–1277.

Ferreira, V. and Lisboa, A. (2019). Innovation and entrepreneurship: From schumpeter to industry 4.0. In *Applied mechanics and materials*. Trans Tech Publications Ltd., 890, 174–180.

Innocenti, N. and Zampi, V. (2019). What does a start-up need to grow? An empirical approach for Italian innovative start-UPS. *International Journal of Entrepreneurial Behavior & Research*, 25(2), 376–393.

Jain, R. and Khandelwal, R. (2020). Dare to defy the challenges of online business. *International Journal of Entrepreneurship and Innovation Management*, 24(4/5), 281.

Johnson, P. (2006). Business models. *Technology, Innovation, Entrepreneurship and Competitive Strategy*, 53–72.

Kacprzyk, J., Fedrizzi, M. and Nurmi, H. (1992). Group decision making and consensus under fuzzy preferences and fuzzy majority. *Fuzzy Sets and Systems*, 49(1), 21–31.

Kuckertz, A., Brändle, L., Gaudig, A., Hinderer, S., Morales Reyes, C. A., Prochotta, A., Steinbrink, K. M. and Berger, E. S. (2020). Startups in times of crisis—A rapid response to the COVID-19 pandemic. *Journal of Business Venturing Insights*, 13, e00169.

Kumar, V., Bharti, I. and K. Sharma, Y. (2012). Fuzzy diagnosis procedure of the types of glaucoma. *International Journal of Applied Information Systems*, 1(6), 42–45.

Kumar V. and Vashist D. (2013). Fuzzy procedure for the selection of car among various brands. *International Journal of Engineering Research and Technology*, 6(3), 337–342

Mehlawat, M. K. and Gupta, P. (2014). Credibility-based fuzzy mathematical programming model for portfolio selection under uncertainty. *International Journal of Information Technology and Decision Making*, 13(1), 101–135.

Müller, J. M., Buliga, O. and Voigt, K. I. (2018). Fortune favors the prepared: How SMEs approach business model innovations in Industry 4.0. *Technological Forecasting and Social Change*, 132, 2–17.

Nagar, D., Raghav, S., Bhardwaj, A., Kumar, R., Lata Singh, P. and Sindhwani, R. (2021). Machine learning: Best way to sustain the supply chain in the era of industry 4.0. *Materials Today: Proceedings*, 47, 3676–3682.

Pangaribuan, I., Zaka, M. and Yunanto, R. (2021). Design of web-based online sales: As an entrepreneurship strategy. *International Journal of Entrepreneurship & Technopreneur*, 1(1), 31–36.

Perry, C. (2000). Case research in marketing. *The Marketing Review*, 1(3), 303–323.

Ponomarev, A. (2019). The five reasons why startups succeed, according to a legendary investor. *Medium*. https://medium.com/swlh/the-five-reasons-why-startups-succeed-according-to-a-legendary-investor-306ff3856e0d

Rezaeinejad, I. and Chernikov, S. U. (2021). Impact of COVID-19 on Iran startups at biotech, pharmaceutical, engineering and other innovative industries. SHS Web of Conferences, 114, 01018.

Robertson, M., Collins, A., Medeira, N. and Slater, J. (2003). Barriers to start-up and their effect on aspirant entrepreneurs. *Education and Training*, 45(6), 308–316.

Salamzadeh, A. and Dana, L. P. (2020). The coronavirus (COVID-19) pandemic: Challenges among Iranian startups. *Journal of Small Business & Entrepreneurship*, 33(5), 489–512.

Sarika, J., Vijay, K. and Arti, S. (2018). Generalized fuzzy information entropy measure: A case study for the selection of diamond among various brands. *Recent Patents on Engineering*, 12(3), 223–229.

Shahab, Y., Chengang, Y., Arbizu, A. D. and Haider, M. J. (2019). Entrepreneurial self-efficacy and intention: Do entrepreneurial creativity and education matter? *International Journal of Entrepreneurial Behavior & Research*, 25(2), 259–280.

Shanker, K., Shankar, R. and Sindhwani, R. (2019). Advances in industrial and production engineering. Select Proceedings of FLAME 2018 Book Series by Springer-Nature.

Shen, H., Fu, M., Pan, H., Yu, Z. and Chen, Y. (2020). The impact of the COVID-19 pandemic on firm performance. *Emerging Markets Finance and Trade*, 56(10), 2213–2230.

Sindhwani, R., Afridi, S., Kumar, A., Banaitis, A., Luthra, S. and Singh, P. L. (2022b). Can industry 5.0 revolutionize the wave of resilience and social value creation? A multi-criteria framework to analyze enablers. *Technology in Society*, 101887.

Sindhwani, R., Hasteer, N., Behl, A., Varshney, A. and Sharma, A. (2022a). Exploring "what,""why" and "how" of resilience in MSME sector: a m-TISM approach. *Benchmarking: An International Journal*, In press (ahead-of-print).

Sindhwani, R., Kumar, R., Behl, A., Singh, P. L., Kumar, A. and Gupta, T. (2021). Modelling enablers of efficiency and sustainability of healthcare: A m-TISM approach. *Benchmarking: An International Journal*, 29(3), 767–792.

Singh, P. L., Sindhwani, R., Sharma, B. P., Srivastava, P., Rajpoot, P. and Kumar, R. (2022). Analyse the critical success factor of green manufacturing for achieving sustainability in automotive sector. In *Recent trends in industrial and production engineering.* Springer, 79–94.

Startups Boom in the United States during COVID-19. (2021). PIIE. www.piie.com/blogs/realtime-economic-issues-watch/startups-boom-united-states-during-covid-19

Szmidt, E. and Kacprzyk, J. (2000). Distances between intuitionistic fuzzy sets. *Fuzzy Sets and Systems*, 114(3), 505–518.

The Case Study Approach (2011). BMC medical research methodology. https://bmcmedresmethodol.biomedcentral.com/articles/10.1186/1471-2288-11-100

Vaznyte, E. and Andries, P. (2018). Entrepreneurial orientation and Start-UPS' external financing. *Academy of Management Proceedings*, 2018(1), 13368.

Yang, X., Sun, S. L. and Zhao, X. (2018). Search and execution: Examining the entrepreneurial cognitions behind the lean startup model. *Small Business Economics*, 52(3), 667–679.

Zadeh, L. A. (1965). Fuzzy sets. *Information and Control*, 8, 338–353.

10 Charting Industry 4.0 Routes Incubation Centers
A Study on Atal Incubation Centre

Kshitiz Choudhary, Jayant Mahajan,
Fr Jossy P. George and Anshul Saxena

CONTENTS

10.1 Introduction.. 153
    10.1.1 Industry 4.0 154
10.2 Start-Up: Concept & Beyond... 154
    10.2.1 Start-Up Incubator 155
    10.2.2 Mission and Purpose of Business Incubators 156
    10.2.3 Incubation Evolution.. 157
    10.2.4 Role of Incubators in Industry 4.0 157
10.3 Atal Incubation Centers (AICs)... 159
    10.3.1 AICs' Expected Functions 161
    10.3.2 AIC's Advantages 162
    10.3.3 Benefits of AICs... 162
    10.3.4 AIC Purpose for Start-Ups.. 163
10.4 Start-Up India: Catalyst of Change... 165
10.5 Established Incubation Centers (EICs) 165
10.6 Discussion.. 166
    10.6.1 How Can Atal Incubation Centre Help Startups? 166
    10.6.2 Findings 169
10.7 Conclusion 170
References... 170

10.1 INTRODUCTION

According to the data on start-ups (Upstream Awards, 2021), there are two character-istics that make India an emerging eco-system for the start-up initiation: a) the cost of doing business which is less due to client proximity as well as merchants and b) the enormous size of the home market, as well as the large number of internet users. When a start-up begins, it must have at least some users who need what it's creating right now not just individuals who could use it in the future, but people who need it right now. This initial set of users is usually limited, for the simple reason that if there was anything that a huge number of people desperately wanted and could be developed with the amount of work a start-up typically invests into a version one, it would almost certainly already exist (Hindu BusinessLine, 2019). Which implies you have to make a trade-off on one dimension: you can either construct something that a huge number of people desire in a little amount, or something that a small number of people want in a large amount. That's how many, if not all, of the world's most successful businesses began. Apple, Yahoo, Google, and Facebook were never intended to be businesses in the first place. Because there seemed to be a void in the world, they developed out of the items their founders produced (AIM, n.d.).

10.1.1   INDUSTRY 4.0

One critical thing to learn from the Covid-19 time in history is that cost-effective new automation systems are essential to keep plants and supply chains functioning efficiently while delivering supreme quality goods, whether OEM, components or assembly supplier, contract manufacturer, or manufacturing services provider. More than 70 terabytes (TB) of big data may be generated on one manufacturing line every day, yet much of this data is still unanalyzed. When pace and communication latency are crucial, hybrid cloud incorporates the newest next-level information technologies like artificial intelligence (AI), edge computing (EC), and cybersecurity to effectively derive wealth and act on this underutilized data (Al-Mubaraki and Busler, 2017).

In several industries, including FMCG, automobile, healthcare, aviation, defense, and many others, the rise of the "smart factory" is being spurred by the advent of Industry 4.0. By using intelligent and linked digital technology, automated systems will be able to further improve their efficiency, profitability, compliance, and satisfaction with their customers. By 2025, the digital manufacturing business is predicted to be worth $767.82 billion, an increase of 7.8% over the previous year (Atal Innovation Mission, n.d.). Manufacturing is an important part of India's digital transformation agenda, as with many other nations. An increasing need for experts in new technologies to spearhead the digital transformation of the industrial industry has emerged.

10.2   START-UP: CONCEPT & BEYOND

A start-up business is one that has just begun operations. Entrepreneurs that want to build a product or service that they believe will be in high demand from start-ups. Venture capitalists and other forms of funding are often sought by these enterprises because of the high expenses and poor profits they normally face at the start. The term "start-up" refers to enterprises or efforts that concentrate primarily on one product or service. As a result, these firms frequently lack a fully established company plan and the financial resources necessary for further growth. The founders of most of these enterprises are the primary source of funding. Seed capital may be used by start-ups to support the research and innovation of their business plans. A thorough

Startup Incubators In India
Geographic Spread

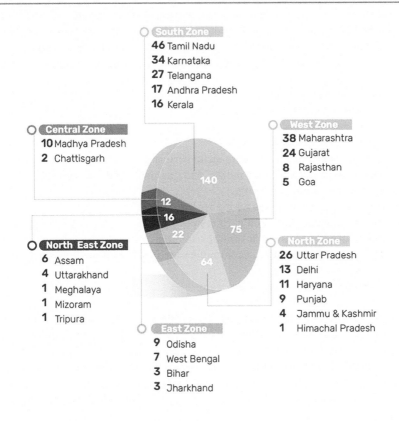

Source: Shifting Orbits: Decoding the Trajectory of the
Indian Start-up Ecosystem

Bloomberg | *Quint*

FIGURE 10.1 Shifting orbits: Decoding the trajectory of the Indian start-up ecosystem (Bloomberg Quint).

business plan includes the firm's stated mission, vision, and objectives, and also management and marketing strategies, while market research helps determine whether or not a product or service is in demand. Figure 10.1 depicts the Indian start-up ecosystem using Bloomberg quint's.

10.2.1 START-UP INCUBATOR

A start-up incubator is a collaboration group that assists new firms in growing. Incubators help entrepreneurs overcome several of the hurdles that come with starting a business by providing workspace, early funding, coaching, and training. The primary goal of a start-up incubator is to support entrepreneurs in expanding respective firms (Business Incubators, n.d.). Start-up incubators are frequently non-profit organizations run by both public and commercial entities. Incubators are typically associated with colleges and business schools.

Incubators have evolved into a global economy with a myriad of services that encourage and enhance economic growth by nurturing the establishment and creation of innovative firms, with astounding outcomes in developed economies.

The following is a list of the most often supplied services by business incubators.

- Help with business principles
- Networking good opportunities
- Marketing assistance
- High bandwidth connectivity
- Management of accounting/financial performance
- There are bank loans, loan money, and guarantee plans accessible
- Help with presentation skills
- Connections to resources for higher learning
- Relationships to important partners
- Relationship with angel or venture capital investors
- Mentorship and consulting committees are institutes providing comprehensive business training
- Formation of the management team

Business incubators help start-ups manage their finances and put their money to the greatest possible use. Scaling up a firm begins with a solid foundation in the local market. Business incubators essentially serve the same purpose. They target businesses that want to formally establish themselves in the market. Such businesses with high growth potential may require various types of assistance, such as planning, training and development, research assistance, and so on, which business incubators provide.

A business incubator creates a supportive environment that helps you build business ideas and turn concepts into marketable products, assist you in acquiring business knowledge, assist you in raising the necessary funding, and educate entrepreneurs on business networks. It greatly reduces the risk of failure. They not only cut costs, making it easier for aspiring entrepreneurs to start a business but also increase their chances of survival and success by expanding their opportunities and networks. A lot of research must be done before a business incubator decides to help or finance a start-up. It is the main purpose of business incubators to develop employment opportunities and commercialize new technology.

10.2.2 MISSION AND PURPOSE OF BUSINESS INCUBATORS

Entrepreneurial enterprises go through many stages throughout their life cycle. They might be in the start-up, business development, or maturity stages (Isabelle, 2013) incubators are most effective when their mission and goals align with the requirements of entrepreneurs as well as sponsoring businesses. They might be in the early phases of development, corporate growth, or maturity (Isabelle, 2013). Incubators are most effective when their mission and goals are aligned with the needs of entrepreneurs and sponsoring corporations (Isabelle, 2013). Moreover, entrepreneurs can analyze the performance measures of the incubator's firm, like the number of customers, customer survival rate, utilization rate, managerial effectiveness, licensing, and funds raised (Isabelle, 2013). When deciding whether to join incubators, entrepreneurs should examine the reputation of the incubator organization since it affects the exposure of the entrepreneurial enterprise as well as its capacity to attract funds, assets, and talent (Isabelle, 2013).

10.2.3   Incubation Evolution

To restate, the idea of incubation, as well as its function, services, and outcomes, has evolved. (Lamine et al., 2015) divided the support ecosystem for new firms into three waves: before the 1980s, 1980–1990s, and 2000–2014. The first wave is driven by research parks and technology development centers, while the second wave focuses on commercialization and adoption of mentoring and networking. The third wave that emerged in the 2000s emphasized the creation of models such as specialized incubators and accelerators. Although incubation has been historically defined regardless of industry or mainstream, current research focuses primarily on technology-based business incubation. In this fast-changing era, two standout models are getting a lot of attention. Technology Business Incubators, or TBIs, and accelerators. While incubation has historically been defined without regard to industry or emphasis areas, current research has focused mostly on technology business incubation. In these rapidly changing scenarios, two prominent frameworks receive a lot of attention: Technology Business Incubators, or TBIs, and accelerators. Accelerators are the latest addition to our startup support plan. Unlike incubators, which are physical "corporations," accelerators are fixed-term "programs" that provide input to participating entrepreneurs by providing them with training and equipment to address their challenges, thereby introducing them to investors and preparing them to raise funds. The first accelerator was the US YCombinator, founded in 2005. In less than 9 years from 2005 to 2013, more than 213 accelerators worldwide supported more than 3,800 new e-incubator services and activities.

Data was collected through an online survey from some incubators done by Atal Innovation Mission connected with one of the schemes mentioned in the preceding section in order to gain a better knowledge of their offerings, activities, experiences, and difficulties. With equivalent goals and resources for all incubators, it was assumed that any disparities in their success could be attributed to changes in their characteristics (such as geography) or other internal business procedures.

10.2.4   Role of Incubators in Industry 4.0

Small and medium enterprises (SMEs) are important for balanced socio-economic development in both established and developing nations, and their growth is aided by a variety of policies, as mentioned above. In both established and emerging countries, the failure rate of tiny new firms in their first years is significant. This is due in part to the competitive climate in which the enterprises are established, as well as the viability of the individual company innovation. It is also a result of the lack of experience of entrepreneurs and the poor environment when setting up a company (lack of capital, legal difficulties, lack of information, etc.). The government supports a variety of initiatives to reduce bankruptcy rates by addressing environmental challenges (funding special loans, removing legal barriers, reducing government administrative processes, and speeding up operations) and helping new entrepreneurs overcome their lack of experience.

Business incubators not only provide support for entrepreneurs through a supportive environment that helps them realize business ideas and transform concepts into industrial products, but also assist in acquiring business knowledge, raising necessary funds, and connecting to corporate networks., all should contribute to minimizing the failure rate. They not only make it easier for start-ups to start a company by reducing the associated costs and risks, but also increase their chances of survival and success by expanding their capabilities and connections.

Governments around the world support incubators in a variety of ways. Governments view incubators as powerful tools to support the growth of SMEs and meet diverse socio-economic needs such as job creation, technological innovation transfer and consequent competitive advantage, regional development and restructuring, poverty alleviation and inclusion. of economically vulnerable groups. As a result, public policies supporting incubation can have multiple strategic directions, each with a separate business system for small businesses that is reflected in the core technologies and services provided.

Finding start-up ideas is a delicate business, which is why so many individuals who attempt it fail terribly. It's not enough to try to come up with start-up concepts. You'll get awful ones that seem frighteningly believable if you do it. The ideal way is more indirect: if you come from the proper background, fantastic business ideas will come to you naturally. Even then, it won't happen right away. It takes time to encounter situations when something is lacking. Often, these gaps don't appear to be business ideas, but rather something that might be fun to develop. That's why having the time and desire to develop things just because they're fascinating is beneficial.

An incubator is both a location and a program dedicated to assisting new firms and ensuring their development. A team of specialists provides a set of services aimed at assisting entrepreneurs with the issues outlined above. The form of the facilities varies depending on the sort of business being incubated, however they are given to emerging enterprises at a cheaper rate than usual. Lower rents are an important component of an incubator's overall service in assisting new firms to flourish.

An incubator is formed to give financial assistance to deserving start-ups, and prospective new firms are typically required to compete for admittance. This implies that each incubator must be adequately supported in order to transmit manpower and facility benefits to its new business tenants. An incubator is formed to give financial assistance to deserving start-ups, and prospective new firms are typically required to compete for admittance. This implies that each incubator must be adequately supported in order to transmit manpower and facility benefits to its new business tenants.

The most frequent and effective type of incubator is the business incubator (also known as the Technology Business Incubator). The business incubator provides space and services for start-ups that are associated with the essence of producing or using the fruits of technology in diverse ways. These "tech" firms are always evolving and changing, and they must invest their cash in the technology and equipment required to turn their excellent ideas into profitable commercial operations. Incubator buildings for IT industry start-ups are often extremely big, frequently measuring 20,000 square feet or more. The most successful include floor space for primary workstations as well as laboratory space and common areas such as conference rooms, a central copy center, lounge space, and restroom areas.

10.3 ATAL INCUBATION CENTERS (AICS)

The government of India established the Atal innovation mission (AIM) at NITI Aayog (Figure 10.2—List of Affiliating Government Bodies) (Chinchwadkar, 2021). Atal innovation missions intends to aid in the creation of new greenfield incubation centers known as Atal Incubation Centers (AICs), which will nurture innovative start-ups in their pursuit of becoming cost-effective and commercial organizations.

Table 2: Affiliating Government Bodies

Abbreviation	Full Name
AIM	Atal Innovation Mission, NITI Aayog, Government of India
DARE	Department of Agricultural Research and Education, Ministry of Agriculture and Farmers Welfare, Government of India
DBT	Department of Biotechnology, Ministry of Science and Technology, Government of India
DoS	Department of Space, Government of India
DSIR	Department of Scientific and Industrial Research, Ministry of Science and Technology, Government of India
DST	Department of Science & Technology, Ministry of Science and Technology, Government of India
MDoNER	Ministry of Development of North Eastern Region, Government of India
MEITY	Ministry of Electronics and Information Technology, Government of India MoD
MoFPI	Ministry of Food Processing Industries, Government of India
MoSDE	Ministry of Skill Development and Entrepreneurship, Government of India
MoT	Ministry of Tourism, Government of India
MSME	Ministry of Micro, Small and Medium Enterprises, Government of India

FIGURE 10.2 Affiliating government bodies.

Source: Atal Innovation Mission.

AIM will help these AICs in creating world-class incubation facilities and services by offering incubate start-ups with cutting-edge physical facilities in the form of machinery and equipment and functioning facilities, as well as the availability of sectoral experts for mentorship. Aside from that, support with business growth, access to seed capital, industry partnerships, training, and other vital aspects are accessible (Figure 10.3—Criterion to become Atal Incubation Centre).

The primary goal of this Mission is to foster an environment of innovation and entrepreneurship in India (Canvanizer, 2018). The government recognizes the need to establish world-class incubation facilities in different parts of the country, with optimal physical facilities in overall capital appliances and functioning amenities, as well as the provision of sector specific consultants for guiding start-ups, corporate planning assistance, connect directly to seed capital, industry players, training events, as well as other meaningful components needed to inspire innovative start-ups. As

Table 3: AIM's Criteria to Evaluate Incubators*

Quantum Metrics (count of)	Impact Metrics	Financial Metrics
• Startups supported till date	• Number of jobs created per startup each year	• Quantum of seed funding corpus
• Startups graduated/exited till date	• Annual taxes paid by supported startups	• Non-grant revenue, not including any interest on seed fund
• Physically incubated startups	• Number of technologies patented	
• Virtually incubated startups		• Sustainability (without access to any governmental or non-governmental grants)
• Associated academic institutions	• Number of awards received by startups	
• Entrepreneurship development workshops organised annually	• Cumulative sales turnover of graduated startups	
• Training programs organised		
• Active mentors		
• Active industry/corporate partnerships		
• Personnel at the incubation centre		

*the three categories of metrics are a result of the authors' analysis; this is not how it is represented by AIM

FIGURE 10.3   Criterion to become Atal Incubation Center.

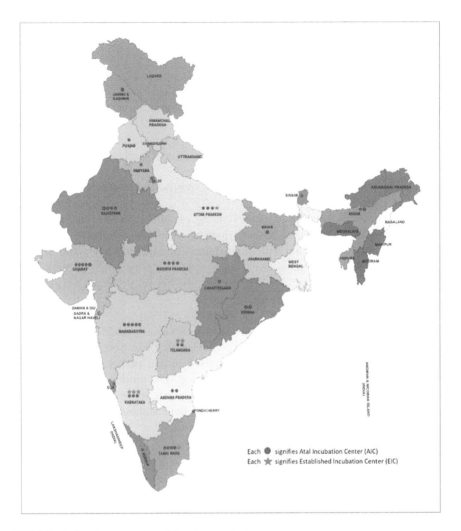

FIGURE 10.4   Coverage of Atal Incubation Center.

a result, AIM argues for the establishment of Atal Incubation Centers (AICs), which would foster innovative start-up businesses on their journey to become suitable and sustainable organizations (Figure 10.4—Coverage of Atal Incubation Centre).

10.3.1   AICs' EXPECTED FUNCTIONS

Expected functions of AIC is covered below and depicted in

- Assist incubators in developing a viable, scalable, and sustainable company strategy.
- Physical infrastructure and value-added support services are provided by providers.

- Build a robust network of mentors who can give sector-specific information as well as real-world practical advice.
- Make inventions into businesses.
- Organize inspiring events and programs.
- Entrepreneurs are given training and coaching.
- Form new and existing relationships and communications with academics, business, sources of financing, and others.
- Start-ups can benefit from incubators and other resources.
- Allow access to prototype facilities, test beds, and markets for the product or service's pilot deployment.
- Create a competent staff with sufficient expertise and experience coaching start-ups.
- Create business strategies, arrange funding, establish networks, and so forth.
- A partnership between a business industry firm and a research-oriented or educational institution with aligned areas of concentration would be an appropriate application.

10.3.2   AIC's Advantages

The incubation method enables entrepreneurs to save cash while gaining external help to accelerate the growth of their firms. The Enterprise Center embraces each entrepreneur's individuality and delivers assistance and personalized services to enhance business potential via business incubation. The ultimate purpose of incubation is to help entrepreneurs create lucrative, long-term businesses. Graduation is chosen collaboratively based on mastery of fundamental business operations; for most businesses, it will occur within four to six years.

- Client advantages
- Assistance with business and technology
- A network of relationships with other business owners affordably/flexibly leased space
- Adaptability of business model to market conditions
- Financial support and counseling
- Capital investment protection

10.3.3   Benefits of AICs

With several advantages of enrolling in an incubator program, small business counsellors regularly urge its clients to investigate the possibility of securing a seat in one. Incubators present the best benefits:

A. **Sharing in basic operating costs:** Utility bills, office furniture and supplies, computer services, and receptionist services are all shared by renters in a business incubator. Furthermore, basic rent prices are often lower than the average in the place where the fledgling business works, allowing entrepreneurs to save significantly. However, It's important to keep in mind that incubators don't enable renters to stay in the program indefinitely; most lease agreements at incubator facilities last three years, with the possibility of one or two one-year renewals.

B. **Administrative and consulting assistance:** Managers and team of workers individuals at incubators might also additionally regularly deliver beneficial endorse and/or statistics on a huge variety of enterprise difficulties, from advertising to enterprise improvement finance. Small enterprise proprietors need to hold in thoughts that the employees in rate of dealing with the incubator program are usually well-versed in lots of components of the enterprise world. They are a precious aid that need to be absolutely tapped.

C. **Versatility of the incubator concept:** One of the primary benefits of incubators is that the concept may be applied to communities of diverse forms, sizes, demographic groups, and sectors. According to Richard Steffens in Planning, "a special strength of an incubator is its capacity to assist enterprises that satisfy specific needs: technology transfer, reviving areas, providing minority jobs, and so on." In many situations, the incubator adopts some of the characteristics of the community in which it is housed. Rural-based incubators, for example, may create businesses based on the agricultural existing in the region. However, whether headquartered in a tiny town in the Midwest or a huge urban location in the West, supporters of incubator programs argue that the community's small business people would benefit would be more knowledgeable about how to develop and run such enterprises than large organizations focused on mass production.

D. **Community's legitimacy:** Many entrepreneurs note that when a startup is enrolled in a startup incubation program, it obtains trust and respectability from both suppliers and consumers. "The fact that a company is registered in an incubator adds value to the due diligence process for potential investors."

E. **Mentoring:** Most incubators and accelerators provide mentoring and advisory services to their members. Having someone or a group of people guiding you is an important part of helping small businesses. As an entrepreneur, you will benefit greatly from the guidance of someone with real experience in your field or business.

F. **Savings in respect of time and money:** You will save time and money if you participate in a business incubation program. It takes time to create a new product or service while also establishing the relevant company structures. As a result, enrolling in a program where you will study, receive assistance, and swiftly bring your business to market makes a lot of sense.

G. **Enhances company growth:** The main goal and benefit of corporate incubation is to accelerate business growth. This is what sets business incubators apart from other types of initiatives. Startups or small businesses participating in incubation programs are most likely to succeed.

10.3.4   AIC PURPOSE FOR START-UPS

The main goal of the incubator is to develop a successful business to make the program financially sustainable and self-sufficient. Graduates of these incubators can create jobs, revitalize communities, commercialize advanced technologies, and revitalize local and national economies. Incubators typically provide clients with adequate rental space and flexible rentals, common core business services and equipment, technical support services, assistance in securing the necessary funds to grow the company, and access to mentors to assist the company in its early stages. Incubators also play an active role in the local economy, serving as a point of contact for venture capitalists, academics, college students, and government officials. This powerful key stakeholder allows for an active exchange of ideas and the development of rapid and effective solutions to challenges that affect emerging enterprises in a nation. Figure 10.5 reports the impact of Atal Incubation Centres and its contribution areas. The advantages of this reach not just to the stakeholders stated above, but also to the nation's overall economy (Figure 10.6—Guiding Principle of Atal Incubation Center).

In addition, the scheme seeks to encourage and build top quality incubation centers in India in certain subjects/sectors such as manufacturing, transportation, energy, health, education, agriculture, water, and sanitation, and so on. These incubation centers would promote and foster creative technology-based start-ups with a potential application and/or influence in the economy's main sectors. The incubation centers would give start-ups with mentoring, technical assistance, infrastructure, access to investors, networking, and a variety of other resources that may be required for the firm to survive and expand.

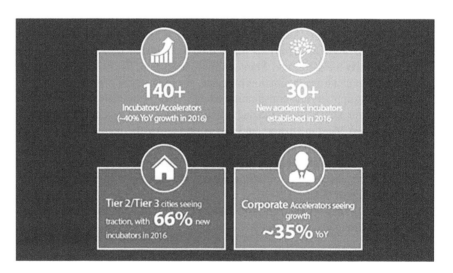

FIGURE 10.5   Impact of Atal Incubation Center.

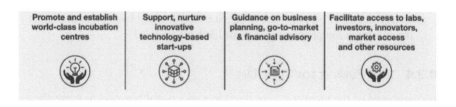

FIGURE 10.6   Guiding principle of Atal Incubation Centre.

10.4 START-UP INDIA: CATALYST OF CHANGE

Start-up India has devised a three-pronged approach to operationalize its action plan toward a "pro-entrepreneurship" attitude:

a. To provide a single platform to link the whole ecosystem while minimizing information asymmetry
b. To offer benefits and other essential assistance
c. To assist area entrepreneurs in turning their ideas into viable commercial businesses

While the first two seek to provide a one-stop shop for entrepreneurs, incorporating financial incentives and other services, the third point's goal is to be reached through outreach and awareness initiatives by taking the extending the message beyond metros to tier 2 and 3 cities (Lamine et al., 2015), and including regional governments in the overall change. The Startup India Initiative's Action Plan was largely focused on tackling the most critical challenges encountered by entrepreneurs at various stages of their firms and growing the Indian (Startup India).

The startup ecosystem to become one of the largest startup ecosystems in the world. Since the introduction of Startup India and the subsequent execution of numerous programs, there has been a significant improvement in the entire startup ecosystem. Furthermore, the Ease of Doing Business index for India has increased significantly from 130 in 2016 to 63 in 2020 (Venugopal, 2017), boosting the growth of start-ups.

10.5 ESTABLISHED INCUBATION CENTERS (EICS)

In recent years, academics, industry, investors, entrepreneurs, government agencies, and non-governmental organizations have taken the lead in establishing incubation centers around the country (CIS, n.d.). These are incubators established for start-ups that need help building connections between incubators, institutions and businesses to expand, improve and update incubator capabilities and create a supportive environment. AIM plans to revitalize the business incubation center (EIC) established in Korea through financial support for scale expansion. This program will radically improve the entrepreneurial environment in the country, raising EIC to a global standard. These EICs will support creative and fast-growing businesses, contributing to the development of the country's vibrant business environment. AIM intends to revitalize the country's Established Incubation Centers (EICs) by providing financial scale-up support.

Educational and non-academic entities (AEG, n.d.), including businesses, innovative investment funds registered with SEBI and corporate accelerators, and individual groups, are eligible to participate in the program. And individuals, and so on. AIM would give a grant-in-aid of up to Rs. 10 crores for a maximum of five years to cover capital and operating expenses for the planned AIC.AIM will make a grant-in-aid of up to Rs. 10 crores available to the applicant institution in two or more annual instalments. The various types of incubators are as follows:

- *Pre-incubators* provide services primarily related to the pre-incubation phase of the incubation process. They provide the experience (coaching and direct support) and equipment (at least a workstation) to help aspiring entrepreneurs develop their company's ideas and plans.
- *Academic incubators* are located in universities and research centers and support companies to develop concepts that students develop or are a by-product of their research activities.
- *General incubators* provide a wide range of services from pre-incubation to post-incubation, as well as support for anyone with a viable concept, regardless of origin or sector of the economy.
- *Sector incubators* provide all services from before to after incubators and help anyone with a viable concept within a particular sector of the economy that is an expression of the true endogenous potential of the region in which the incubator is located.

While providing business services to supported entrepreneurs (RMPSCo, n.d.), enterprise hotels focus primarily on physical incubation activities and are a regular occurrence in major urban regions where manufacturing and office space are a hindrance.

10.6 DISCUSSION

No doubt, the incubators have an impact on entrepreneurship and even help in nurturing the start-ups. That's why the incubators have become an essential mechanism to allow new firms to take birth and flourish for the greater benefit of the economy. However, in order to add research value to this study, an attempt is made to determine whether the presence of a successful incubator within local proximity or within regional news-reach encourages or motivates kids to pursue a career as an entrepreneur.

10.6.1 How Can Atal Incubation Centre Help Startups?

Mentoring

A brilliant concept alone cannot maintain or expand a business. Young entrepreneurs may struggle with a variety of crucial elements ranging from management to market strategy to regulatory backing.

In a knowledge-based economy, start-ups might acquire insights from mentors that would otherwise be difficult to obtain.

Incubators provide access to a panel of mentors who are committed specialists who care about their achievement. Incubators establish a time-bound commitment to see a challenge through, which may be difficult to motivate through an independent mentor.

Incubators provide credibility to entrepreneurs that lack brand recognition/awareness at the outset of their journey, allowing them to increase their client base and get investor access.

Fundraising

In the second week of July, the Indian startup ecosystem attracted $341.3 million in equity funding across 18 transactions (RIT Scholar Works, n.d.). The fact that 85% of this investment was raised by late-stage agreements and less than $12 million by early-stage deals (IIMA, n.d.) underscores the reality that, while finance is increasing, it is not readily available for less established firms. Incubators facilitate access to investors. They provide credibility and aid in the opening of previously closed doors.

While conducting an interview with one of the Incubators, this researcher focused on some critical questions that a Startup should consider, and my observation was in the answer that the Incubator provided, which I thought to be beneficial and focused on as a start-up looking for Incubator.

1. What Are the Incubators?

Business incubators are organizations that help entrepreneurs establish their businesses, particularly in the early stages. These are organizations that are committed to boosting the growth and success of start-ups and early-stage enterprises. Offering technical facilities and advice, seed capital, network and connections, co-working spaces, lab facilities, coaching, and advisory aid are all part of incubation assistance. Incubation is frequently carried out by institutions with commercial and technology knowledge. They are typically a good means of obtaining investment from angel investors, government organizations, economic development coalitions, venture capitalists, and other investors.

2. What are the key points on which a start-up should concentrate during its early stages?

During the early stages of a start-up, the first-time entrepreneur and funders are likely to face a number of issues that are unknowns. The entrepreneur's lack of prior experience creates a knowledge gap, making it difficult for him or her to turn an idea or concept into action reality. These problems and unknowns may be considered as a void that, with the right know-how, may be filled.

3. Why one should go for incubators especially with AIC?

In a market where 90% of startups fail, incubators and accelerators may play a critical role in increasing the number of long-term firms. This type of assistance might be provided through an incubator, which is a business that fosters technological and knowledge-based startups driven business concepts by offering work space, shared office services, and specialized services equipment as well as value-added services such as fundraising and legal services for incorporation or registration, business planning, technical support, and, most significantly, networking aid in terms of assisting them in obtaining their first few consumers.

4. How Do Atal Incubation Centre Help in Economic Growth?

The startup India movement has shone a welcome light on raising awareness and spreading information, but the path ahead is lengthy. Currently, just 8% of incubators in India receive government funding. State-backed incubators in Gujarat, Telangana,

Rajasthan, Kerala, and Karnataka show that government backing is boosting the business environment.

Developing the incubator ecology, which is strongly reliant on the pillars of seed funding, institutions, corporates, and so on, as well as aiming for quality alongside volume, must be prioritized since incubators contain tremendous potential to combat unemployment and poverty, to move the economy ahead by reducing the high failure rate of SMEs, and to foster an environment of innovation and entrepreneurship

5. What are the government's policy measures in this regard?

To create a world-class startup ecosystem in India, the Government of India's NITI AYOG has implemented a slew of policy measures in collaboration with the Departments of Science and Technology, Entrepreneurship and Skill Development, Micro, Small and Medium Enterprises, Industrial Policy and Promotion, and Human Resource Development.

A handful of states have also implemented state-level starting regulations. The scope of this research restricts this study to discussing only two key government schemes: the first is NitiAyog's "Atal Innovation Mission," and the second is the Department of Science and Technology's Grant (DST).

6. Could you shed some insight on how the incubators are organized?

The inclusion of business leaders and university officials on the incubator consulting council is critical. A network of relationships with industry must be established, using concepts such as entrepreneur groups and engagement with business consultants.

- Selecting chances for company formation.
- Evaluating business offers in order to compare and contrast them with others.
- A well-functioning system for tracking and analyzing a company's progress from inception to exit.

7. Can you talk more about the infrastructure given by AIC to its incubatees?

Work is frequently produced by technological start-ups in the form of highly flexible teams of people working together. As a result, the working atmosphere is fairly open and flexible. The building has an open design that allows for modifications in utilities and technological infrastructure, as well as office space arrangement. High-tech laboratories, bottle washers, autoclaves, bathrooms, and conference space are often established in convenient locations; nonetheless, it is preferable if the facility provides a level of incompleteness throughout all incubator platforms.

8. What is its track record of client success?

This question is about the overall momentum of the program's capabilities for its startup clients. Has the incubator had any significant successes in terms of company development, employment creation, or economic impact? How many businesses does it now assist, and how many has it supported to date? What is the success rate of new businesses? You may also wish to poll "insiders" in your entrepreneurial environment to measure external impressions.

9. *Can you tell me about the many types of incubators that are part of the Atal innovation centre program?*

Yes, there are several types of incubators, such as preincubators, which normally provide services linked to the preincubation phase of incubation. Sector specific incubators offer a full range of services from the pre to post incubation phases, as well as assistance to anybody with a viable concept within a specific economic sector, which is a manifestation of the genuine endogenous potential of the region where the incubator is located.

Enterprise hotels, which provide business services to supported entrepreneurs, focus mostly on physical incubation activities and are a typical sight in big urban regions where manufacturing and office space are limited.

10.6.2 FINDINGS

Incubators should focus on four areas, according to the conclusions of the study.

1) To be dynamic models of self-sustaining, efficient, and productive development.
2) To provide useful tools for job creation.
3) To promote and support ventures and innovative thinking to create the best opportunities for business start-ups and smart growth.
4) To encourage value-added businesses through various means, such as creating the region's scientific parks and R&D centers, fostering coordination.

On a larger scale, it was unearthed that India's entrepreneurial and innovation ecosystem is largely directed by government schemes (AJOL, n.d.), or that the government plays a significant role in initiating and enabling the incubation process through seed funding, start-up policies, and assistance with commercializing and patenting new technologies. As per the analysis, government assistance has an effect on early-stage finance, but second-stage funding needs further attention (Inc.com, n.d.).

Furthermore, in addition to financial entities, policy factors have a significant influence on fostering incubation operations, particularly in the creation and commercialization of inventions by incubates. During the course of the study, when analyzing trends in technology adoption, we discovered that rising and high-tech industries such as IT and its related disciplines, biotechnology, pharmaceuticals, and the manufacturing sector have increased dramatically (ResearchGate, n.d.).

Incubators catalyze economic growth and assist vulnerable businesses founded and conceptualized mostly by first-generation entrepreneurs. They enable new technologies and research originating in academic institutions and research laboratories to be commercialized, therefore generating economic and societal value. The collaboration between the government and academic institutions to nurture entrepreneurs through the incubator is mutually beneficial. Most significantly, in order for incubation to be effective and have a greater impact, competent and driven incubator workers must be onboarded (SBA, n.d.), as well as meaningful performance evaluation measures developed.

10.7   CONCLUSION

The case study demonstrates the impact of government financing on initial-stage funding, but second-stage funding requires more focus. Furthermore, in addition to financing entities, policy factors have a significant influence on supporting incubation operations, particularly in the creation and commercialization of innovations by incubates. During the course of the study, when analyzing trends in technology commercialization, we discovered that growing and high-tech industries such as IT and its associated disciplines, biotechnology, pharmaceuticals, and the manufacturing sector have expanded considerably.

Business incubators provide entrepreneurs with a helpful environment that helps them develop business ideas, transform concepts into market-ready product lines, continue to support business knowledge generation, raise necessary funding, and engage entrepreneurs in business networks. We provide tailored support. All of this should significantly reduce the failure rate. It not only allows new entrepreneurs to start their own business while reducing the associated costs and risks, but also increases their chances of survival and success by expanding their opportunities and networks.

REFERENCES

AEG. (n.d.). https://aeg.edu.in/2020/10/03/be-informed-about-what-precisely-is-a-startup-incubator/

AIM. (n.d.). Atal Innovation Mission (AIM). https://aim.gov.in/atal-incubation-centres.php.

AJOL. (n.d.). www.ajol.info/index.php/huria/article/download/168086/157490

Al-Mubaraki, H. M. and Busler, M. (2017, July 24). Challenges and opportunities of innovation and incubators as a tool for knowledge-based economy. *Journal of Innovation and Entrepreneurship*. SpringerOpen. https://innovation-entrepreneurship.springeropen.com/articles/10.1186/s13731-017-0075-y.

Atal Innovation Mission. (n.d.). What is an incubator. *Atal Innovation Mission*. https://aim.gov.in/what-is-an-incubator.php.

Business Incubators. (n.d.). Advantage, percentage, benefits, cost, development of incubators, advantages of incubators. www.referenceforbusiness.com/small/Bo-Co/Business-Incubators.html.

Canvanizer. (2018, March 29). How to develop your startup idea. *Canvanizer*. https://canvanizer.com/how-to-use/how-to-develop-your-startup-idea.

Chinchwadkar, R. (2021, August 2). The geography of startup incubation in India. *BloombergQuint*. www.bloombergquint.com/opinion/the-geography-of-startup-incubation-in-india.

CIS. (n.d.). https://cis-india.org/internet-governance/blog/technology-business-incubators.pdf

The Hindu BusinessLine. (2019, November 30). AIM is to be a springboard for start-ups. *The Hindu BusinessLine*. www.thehindubusinessline.com/news/education/aim-is-to-be-a-springboard-for-start-ups/article30122069.ece?utm_campaign=amp_article_share&utm_medium=referral&utm_source=whatsapp.com.

IIMA. (n.d.). https://web.iima.ac.in/assets/snippets/workingpaperpdf/16280815672020-03-01.pdf

Isabelle, D. (2013). Key factors affecting a technology entrepreneur's choice of incubator or accelerator. *Technology Innovation Management Review*, 16–22.

Lamine, W., Jack, S., Fayolle, A. and Chabaud, D. (2015). One step beyond? Towards a process view of social networks in entrepreneurship. *Entrepreneurship & Regional Development*, 27(7–8), 413–429.

ResearchGate. (n.d.). www.researchgate.net/publication/280836906_The_relevance_and_challenges_of_business_incubators_that_support_survivalist_entrepreneurs

RIT Scholar Works. (n.d.). https://scholarworks.rit.edu/cgi/viewcontent.cgi?article=11380&context=theses

RMPSCo. (n.d.). https://rmpsco.com/2021/05/08/startup-incubation-programme/

SBA. (n.d.). www.sba.gov/sites/default/files/rs425-Innovation-Accelerators-Report-FINAL.pdf

Upstream Awards. (2021, December 20). 11 top benefits of business incubation. *Upstream Awards*. www.upstreamawards.com/benefits-of-business-incubation/.

Venugopal, P. (2017). Incubating enterprises: A case study of THub introduction. www.researchgate.net/publication/325346929_Incubating_Enterprises_A_case_study_of_T-Hub_Introduction

11 Imperatives Associated with Women's Participation in Entrepreneurial Activities

*Om Prakash Mishra, Rajeev Saha
and Rajender Kumar*

CONTENTS

11.1 Introduction...173
11.2 Key Factors of Women Entrepreneurship 174
    11.2.1   Education Opportunities 175
    11.2.2   Sense of Independence 175
    11.2.3   Cooperation and Support by the Agencies/Institutions 176
    11.2.4   Opportunities in Contributing the Top Management Decision
             Processes 176
    11.2.5   Increase in Collateral Property 176
    11.2.6   Increased Rate of Self-Employment...176
    11.2.7   Conducive Environment 177
11.3 Challenges Faced by Women Entrepreneurs 177
    11.3.1   Stereotypes on Capabilities..177
    11.3.2   Patriarchal Construct..178
    11.3.3   Familial Constraints 178
    11.3.4   Constraints Related to Funds 178
    11.3.5   Lack of Female Mentors 178
    11.3.6   Lack of Entrepreneurial Aptitude 179
    11.3.7   Exploitation by Middle Men...179
    11.3.8   Understanding Market Trends 179
11.4 Conclusion 179
References...180

11.1 INTRODUCTION

It is not an exaggeration to say that to be the entrepreneur, one has to play several roles, which is not an easy task. An entrepreneur has to initiate, organize, and then operate the business activities in such a manner that he/she would be able to survive and

DOI: 10.1201/9781003256663-11

sustain the enterprise for a long period of services (Nagar et al., 2021). The under-standing of an opportunity, confidence, creativity, and innovativeness are some of the important skill-sets of a good entrepreneur (Foss et al., 2013; Dutta et al., 2022). In present scenario, most women desire to be economically independent through initiating the entrepreneurial ventures (Bullough et al., 2019; Sindhwani et al., 2022a). Earlier, the female community was confined to its main role i.e., taking care of family members (Bjerke and Karlsson, 2013). Now females are not constrained to their role and they are extending their roles from the kitchen activities.

Further, the literature on decision-making at the top level reveals the importance and sensitivity of the decision on firm performance. It is very crucial to take the decision without looking on the future reactions (Birkel et al., 2019; Sindhwani et al., 2022b). In the past few years, the participation of female candidates in top management has increased exponentially. Recently, the present government in India initiated the entry program for Indian women in the defense field. The participation of women in defense and aviation sectors is increased which was male dominated in earlier decades. It is a clear sign that women are at the rising verge in all fields.

In the beginning of 20th century, very few women entrepreneurs were started their business ventures mainly associated with the food and education related businesses. Also, few have started the businesses related to consultancy and services in hospitality, catering, consultants, and public relations etc. (Dutta et al., 2021; Sindhwani et al., 2021). The various reports on entrepreneurial studies reveals the contribution and importance of women's entrepreneurship in promoting the economic growth of developed economies. The studies supported by general surveys which states that around 5% of women in low- and middle-income countries are entrepreneur whereas 35% are aspire to be an entrepreneur (Agarwal et al., 2020). The stats reveals that the women entrepreneurship needs to be strengthen. In this regard, many government agencies are already starting to promote women entrepreneurship at larger scale. The initiatives such as gender equality/biasness are very popular to bridge the gap. The formal belief behind these initiatives is offering the equal opportunities to women entrepreneur to initiate and run the business in effective manner (Ahl, 2006; Shanker et al., 2019).

India's has projected economy $5 trillion economy by 2025, in such scenario entrepreneurship by women must play a bigger role in its economic development (Singh et al., 2022). The data pertaining start-up related information reveals that around 5% of start-ups are founded by women. The higher female ownership of local businesses in related industries predicts greater relative female entry rates. This paper investigates the factors responsible for rising trend of women entrepreneurship around the world. In addition, the challenges before the women entrepreneurs are also explored. The present study has been carried out based upon facts and finding from the available literature.

11.2   KEY FACTORS OF WOMEN ENTREPRENEURSHIP

The promotion of educational sources to women increased rapidly and results in increase of sense of independence, looking for more carrier opportunities and even in innovations. The rising capability of women can be seen globally (Coleman and Robb, 2016). The presence of women heading multi-national companies is rising and shows good leadership quality (Tambunan, 2009). Figure 11.1 reveals the factors identified

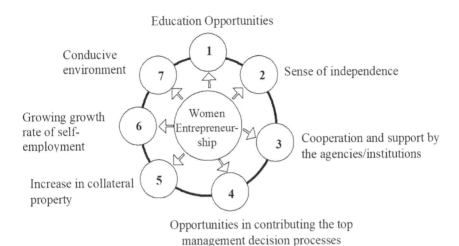

FIGURE 11.1 Factors enabling women entrepreneurship.

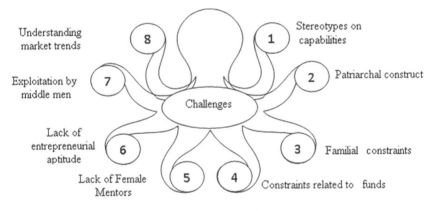

FIGURE 11.2 Challenges for women entrepreneurship.

from the earlier studies on participation of women in entrepreneurship. In addition to this Figure 11.2 illustrates the Challenges faced by women entrepreneurs.

11.2.1 EDUCATION OPPORTUNITIES

Education is right of every one. Earlier, the male domination bound women so they could not get their right for long time especially in developing nations. Women account for two-thirds of the world's illiterate adults, because older women are less likely to have attended school than their younger counterparts. They are also much more likely to be illiterate if they are poor and live in rural areas (Hackler et al., 2008). But the scenario has changed and the involvement of women in education is increasing day-by-day. This leads to changing the perception of entrepreneurship and educated women are making their best decision to innovate their ideas.

11.2.2 SENSE OF INDEPENDENCE

Women are always striving from independence in making decisions regarding their future (Rao, 1991). Now, the social freedom provides an opportunity to them to come forward and make decisions that perfectly suit their personality and capabilities. The women are now free to make decisions related to their personal and professional life (Wahid et al., 2021). Another fact behind this is the acceptance of women's viewpoints as they are more educated and trained in performing the tasks. In addition, there is awareness through campaigns and initiatives by institutions/organizations for empowering the women's help in strengthening their contribution towards social development.

11.2.3 COOPERATION AND SUPPORT BY THE AGENCIES/INSTITUTIONS

Globally, various schemes are already in operation (initiated by the various financial and legal firms) to extend the role and responsibility of women entrepreneurs in the industrial context. The important attribute among cooperation and support is training women for the initiation of their entrepreneurial journey. These kinds of trainings are not only helping them in understanding entrepreneurship but also includes managing the challenges associated. Apart from trainings, the institution promotes the women entrepreneurs through creating platforms where they are guided to build the strong network to add value in the existing businesses or can start from zero (Tyagi et al., 2022). In India, many institutions like DST, EDII, NITI Aayog, NSTEDB promote women entrepreneurship through launching various initiatives and help them in providing an ecosystem for budding women entrepreneurs across the country (Kumari, 2014). The soft loans from various financial institutions supports potential women to turn them into entrepreneurs (Bjerke and Karlsson, 2013).

11.2.4 OPPORTUNITIES IN CONTRIBUTING THE TOP MANAGEMENT DECISION PROCESSES

In businesses, decision-making is the toughest task as it is based on the facts and facets. In addition, strategic planning is a must before making decisions especially for long run perspective. The existing literature on role of women in society reveals that their role is more important within the home boundaries then in the offices (Singh et al., 2022). It was the firm belief of society that a woman was a good manager to manage the home instead of a workplace (industry). Now, the paradigm has shifted and women are also contributing in businesses as well. There is even a rise in women's participation in strategic decision-making. Present-day women employees are getting equal opportunities to come and join the management boards (Rashid et al., 2015). The increased in participation in the leadership decisions help them and motivate others to do the same. This will further lead to rising the entrepreneurial skills among them.

11.2.5 INCREASE IN COLLATERAL PROPERTY

For finance related requirements, in most of the cases collateral was required to meet with the standard requirements of the financing institutions/organizations. Until the 20th century, it is reported that most women entrepreneurs were eligible to get credit from institutions. The biggest reason was non-availability of collateral. Now, the government of developing nations are looking at the fact and initiated many schemes to give benefits to women (Jennings and Brush, 2013). The result of this is a sharp rise in the last decades in having collateral property in women's names.

11.2.6 INCREASED RATE OF SELF-EMPLOYMENT

From ancient times, most women were happy to get themselves involved in household activities rather than professionals. Very few women crossed the house-boundaries and looked for employability to secure their future. The data on women employment worldwide reveals that there is an exponential increase in women employees in last two decades especially in the private sector. Women are now looking for self-employment like men. There are several factors behind this like: greater earnings, giving jobs rather than seeking, exploring environment and trends, own workplace, flexible timing etc. (Fatoki, 2014).

11.2.7 CONDUCIVE ENVIRONMENT

The environment of women empowerment has changed in last two decades. Women are being given due importance in workforce. Such situations are arising as there is a growing demand to maintain gender parity in workforce as well in management (Jamir, 2014; Xie and Lv, 2016). Even, the catering to the customized product demands by the customers provides an opportunity to the women to participate effectively in the businesses. Such encouragement has made women think differently and are excitedly switching towards their own business (Tyagi et al., 2022).

Summary

This section of the study discussed the factors that support the participation of women in entrepreneurship. The above-mentioned factors not only motivate women to go with the entrepreneurial journey but also help them in achieving success.

11.3 CHALLENGES FACED BY WOMEN ENTREPRENEURS

Of course, many schemes and initiatives are being taken by the world community in enhancing the number of women entrepreneurs but some root level problems are still existing in our society. Some of in brief are as followings.

11.3.1 STEREOTYPES ON CAPABILITIES

The stereotypes in societies results in socially structured roles based on gender and creates gender disparity. Even, the current ecosystem of entrepreneurship is more focused on tech-based entrepreneurship solving the needs of the common people (Marlow and Patton, 2005; Kumar et al., 2021). The relationship between entrepreneurial activities & educational condition of women has influenced the participation rate of women in taking leadership roles (Bullough and Abdelzaher, 2013). This needs to identify the various determining factors behind the development of entrepreneurship culture and motivate women to venture into this area.

11.3.2 PATRIARCHAL CONSTRUCT

Though, it is well-known fact about women that they have characteristics such as holistic, inclusive, consultative, and collaborative etc. much better than men. But it is exaggeration to say here that society is still dominated by male members (Doshi, R. and Takalkar, 2016). The regressive belief regarding the woman entrepreneur i.e., they are weak and unstable in managing the activities especially related to businesses is the biggest challenge.

11.3.3 FAMILIAL CONSTRAINTS

No doubt, women are the best managers and manage the requirement of family members in an effective and efficient manner. But while discussing business ventures, the management of activities related to business operations in longer term is much more difficult than managing the family requirements (Kaushik, 2013). In the business operations, the number of resources (such as material, money, manpower, etc.) are required to perform the activities and even, all such activities are having inter-dependencies with each other. Here, one has to look especially at women who are always looking for their family support to meet all the requirements of business such as timely decisions, resource acquisition and accumulation, engagement of activities etc. (Kumar, 2004, Coleman and Robb, 2016). This is often not possible or suspicious especially in the women entrepreneur cases.

11.3.4 CONSTRAINTS RELATED TO FUNDS

For running any kind of business, money plays an important role in the flow of resources from one point to other. It is just like blood moving in human body to keep every part of the body healthier and sensitive t the environment. Though, the count of sources for funding to the entrepreneurial ventures are still increased, there is challenge before women entrepreneurs to take advantage of various sources (Xie and Lv, 2016). There might be many reasons like accessibility, awareness, personal attitude, or lack of family support etc.

11.3.5 LACK OF FEMALE MENTORS

Medium and small-scale industries are being recognized as the engine for economic growth around the world. These industries generally contribute in majority to economy and creates the most noteworthy paces of employment growth. Here, mentors play an important role in initiation of any business venture and their consultation will lead an enterprise towards success. In comparison with male mentors associated with counselling and consultation business for entrepreneurial guidance, very few women have joined this profession (Jain, 2015). In Asia, it is a challenge before the entrepreneurial contribution by women entrepreneurs because it has been observed that women are feeling safe and secure when they are mentored by women mentors as compared to men mentors (Rashid et al., 2015).

11.3.6   Lack of Entrepreneurial Aptitude

Entrepreneurship is not just about starting with a new enterprise, it begins with an idea, attitude to convert that idea into reality and continue until an enterprise stops functioning completely. This means that entrepreneurial activity can be seen in a wide variety of companies, regardless of size, age, and even profit-oriented (Horvath and Szabo, 2019). The stats on entrepreneurship reveal that male entrepreneurs persist and pursue effectively. Even, only few women are revealing the same kind of persistence and pursuance for an opportunity (Wahid et al., 2021).

11.3.7   Exploitation by Middle Men

In entrepreneurial ventures, the opportunity is encashed through strategically to build the arrangement of assets for smooth functioning. For the efficient strategies, the number of stake-holders plays an important role like, suppliers at different tiers, consumers, representatives of financial/regulatory institutions etc. (Wahid et al., 2021). For women entrepreneurs, it is difficult to develop a platform where all stake-holders come and contribute according the desires. This also creates fear among women entrepreneurs to go forward for the entrepreneurial journey (Hackler et al., 2008).

11.3.8   Understanding Market Trends

In the present scenario, due to the emergence of the start-up eco-system, the market becomes more diversified and complex. This result in fierce competition and every industry is now striving to face that one. It becomes more difficult when the business is run by a woman. Behind this, there are several personal and professional related issues which need to be addressed systematically and strategically. Most women are firm believers and spiritual and they always go with the belief system where as the market understanding requires strong desires to compete which is the probable cause that very few women entrepreneurs are successful to date.

Summary

This section of the study discussed the challenges before women to participate in entrepreneurial growth. The challenges discussed above partially bound the women to think about the entrepreneurial journey.

11.4   CONCLUSION

Besides caring the home and family, women are now equally accelerating themselves for self-employment. The desires among them to build wealth for themselves and support to the family, they are capitalizing their capabilities through owning the businesses. Strong advocacy in the current scenario by the society appealing for developing the culture where women can initiate business and compete in male dominated markets. The present study was based on the imperatives of women's participation. The study reveals the seven factors that motivate women to go for entrepreneurial journey. In addition, these factors strengthen their journey throughout and provide them better experiences than ever before.

The study also reveals the challenges before women to go with entrepreneurial journey. Here, it becomes pertinent to point out the role of society towards entrepreneurship which reveals that the challenges stated above are very important to eliminate gender disparity and increased in nation economy. The right kind of education/skilling to women's result in increase of their participation and further reveals their contribution in industry growth perspective. The discussed factors and challenges in the study are clearly depicts women participation i.e., their importance to be an entrepreneur and the hurdles in the path to success. Today, women are having the potential the potential and determination to setup, uphold and supervise their own enterprise in a very systematic manner, appropriate support and encouragement from the society, family, government can make these women entrepreneur a part of mainstream of national economy and they can contribute to the economy progress of India. The most preferred sector for women entrepreneurs is social areas and sectors such as health, education, women's hygiene, fashion, cosmetics, food and nutrition, garments and textiles. Even some of the service sectors such as management of human resources, nursing and homecare are exclusively dominated by women entrepreneurs.

Despite having the fruitful outcomes of the study in terms of imperatives, the present study has limitations also. The study is based on the review pattern and very few imperatives are taken to make this study general than the specific geographic considerations. There might be the chance where any important imperative may not be covered. In addition, the study is not covering the ways to overcome the challenges before women candidates who are willing to go for entrepreneurial journey.

REFERENCES

Agarwal, S., Lenka, U., Singh, K., Agrawal, V. and Agrawal, A. M. (2020). A qualitative approach towards crucial factors for sustainable development of women social entrepreneurship: Indian cases. *Journal of Cleaner Production*, 274, 123135.

Ahl, H. (2006). Why research on women entrepreneurs needs new directions. *Entrepreneurship Theory and Practice,* 30(5), 595–621.

Birkel, H. S., Veile, J. W., Müller, J. M., Hartmann, E. and Voigt, K. I. (2019). Development of a risk framework for Industry 4.0 in the context of sustainability for established manufacturers. *Sustainability*, 11(2), 384.

Bjerke, B. and Karlsson, M. (2013). Women and social entrepreneurship—a comment. In *Social entrepreneurship*. Edward Elgar, 160–164.

Bullough, A. and Abdelzaher, D. (2013). Global research on women's entrepreneurship: An overview of available data sources and limitations. *Business and Management Research,* 2(3), 42–58.

Bullough, A., Hechavarría, D. M., Brush, C. G. and Edelman, L. F. (2019). *High-growth women's entrepreneurship: Programs, policies and practices*. Edward Elgar Publishing, 1–11.

Coleman, S. and Robb, A. (2016). Financing high growth women-owned enterprises: Evidence from the United States. In *Women's entrepreneurship in global and local contexts*. Edward Elgar Publishing, 183–202.

Doshi, R. and Takalkar, S. D. (2016). A study of challenges faced by women entrepreneurs in India. *KRSCMS Journal of Management,* 6(6), 84–91.

Dutta, G., Kumar, R., Sindhwani, R. and Singh, R. K. (2021). Digitalization priorities of quality control processes for SMEs: A conceptual study in perspective of Industry 4.0 adoption. *Journal of Intelligent Manufacturing*, 32(6), 1679–1698.

Dutta, G., Kumar, R., Sindhwani, R. and Singh, R. K. (2022). Overcoming the barriers of effective implementation of manufacturing execution system in pursuit of smart manufacturing in SMEs. *Procedia Computer Science*, 200, 820–832.

Fatoki, O. (2014). Factors motivating young South African women to become entrepreneurs. *Mediterranean Journal of Social Sciences*, 5(16), 184.

Foss, L., Woll, K. and Moilanen, M. (2013). Creativity and implementations of new ideas. *International Journal of Gender and Entrepreneurship,* 5(3), 298–322.

Hackler, D., Harpel, E. and Mayer, H. (2008). Human capital and women's business ownership. In *Office of advocacy U.S. small business administration*. 1–71.

Horvath, D. and Szabo, R. Z. (2019). Driving forces and barriers of industry 4.0: Do multinational and small and medium-sized companies have equal opportunities? *Technological Forecasting and Social Change*, 146, 119–132.

Jain, S. (2015). Challenges faced by women entrepreneurs in India. *Research Journal of Humanities and Social Sciences,* 6(3), 213.

Jamir, T. (2014). Motivational and environmental factors of women entrepreneurs. *Asian Journal of Research in Business Economics and Management*, 4(10), 79.

Jennings, J. E. and Brush, C. G. (2013). Research on women entrepreneurs: Challenges to (and from) the broader entrepreneurship literature? *Academy of Management Annals,* 7(1), 663–715.

Kaushik, S. (2013). Challenges faced by women entrepreneurs in India. *International Journal of Management and Social Sciences Research*, 2(2), 6–8.

Kumar, A. (2004). Enterprise location: Choice of women entrepreneurs. *Small Enterprises Development, Management and Extension Journal (SEDME): A worldwide window on MSME Studies,* 31(3), 11–20.

Kumar, R., Sindhwani, R., Arora, R. and Singh, P. L. (2021). Developing the structural model for barriers associated with CSR using ISM to help create brand image in the manufacturing industry. *International Journal of Advanced Operations Management*, 13(3), 312–330.

Kumari, N. (2014). *Women entrepreneurship in India: Understanding the role of NGOs*. Notion Press, 27–51.

Marlow, S. and Patton, D. (2005). All credit to men? Entrepreneurship, finance, and gender. *Entrepreneurship Theory and Practice*, 29(6), 717–735.

Nagar, D., Raghav, S., Bhardwaj, A. Kumar, R., Singh, P. L. and Sindhwani, R. (2021). Machine learning: Best way to sustain the supply chain in the era of industry 4.0. *Materials Today: Proceedings*, 47(13), 3676–3682

Rao, C. H. (1991). Promotion of women entrepreneurship: A brief comment. *SEDME (Small Enterprises Development, Management and Extension Journal)*, 18(2), 21–27.

Rashid, K. M., Ngah, H. C., Mohamed, Z. and Mansor, N. (2015). Success factors among women entrepreneur in Malaysia. *International Academic Research Journal of Business and Technology*, 1(2), 28–36.

Shanker, K., Shankar, R. and Sindhwani, R. (2019). Advances in industrial and production engineering. In *Select proceedings of FLAME 2018 book series*. Springer-Nature.

Sindhwani, R., Afridi, S., Kumar, A., Banaitis, A., Luthra, S. and Singh, P. L. (2022b). Can industry 5.0 revolutionize the wave of resilience and social value creation? A multi-criteria framework to analyze enablers. *Technology in Society*, 101887.

Sindhwani, R., Hasteer, N., Behl, A., Varshney, A. and Sharma, A. (2022a). Exploring "what,""why" and "how" of resilience in MSME sector: A m-TISM approach. *Benchmarking: An International Journal*, In press (ahead-of-print).

Sindhwani, R., Kumar, R., Behl, A., Singh, P. L., Kumar, A. and Gupta, T. (2021). Modelling enablers of efficiency and sustainability of healthcare: A m-TISM approach. *Benchmarking: An International Journal*, 29(3), 767–792.

Singh, P. L., Sindhwani, R., Sharma, B. P., Srivastava, P., Rajpoot, P. and Kumar, R. (2022). Analyse the critical success factor of green manufacturing for achieving sustainability in automotive sector. In *Recent trends in industrial and production engineering.* Springer, 79–94.

Tambunan, T. T. (2009). Development constraints. In *SMEs in Asian developing countries.* Springer, 159–184.

Tyagi, V., Kumar, R., Singh, P. L. and Shakkarwal, P. (2022). Barriers in designing and developing products by additive manufacturing for bio-mechanics systems in healthcare sector. *Materials Today: Proceedings*, 50(5), 1123–1128.

Wahid, A. N., Aziz, A. N. N., Ishak, M. and Hussin, A. (2021). The critical success factors of business growth among women entrepreneurs in Malaysia: A qualitative approach. *International Journal of Academic Research in Business and Social Sciences,* 11(9), 1445–1459.

Xie, X. and Lv, J. (2016). Social networks of female tech-entrepreneurs and new venture performance: The moderating effects of entrepreneurial alertness and gender discrimination. *International Entrepreneurship and Management Journal,* 12(4), 963–983.

Index

A

angel investing, 98
antecedents of social capital, 56
    community participation, 59
    pro-social motivation, 59
    sense of belongingness, 58
    social identity, 58
artificial intelligence (AI), 111
Atal Incubation Centers (AICs), 159, 160, 166
automated manufacturing, 112

B

big data (BD), 109
business model, 7

C

cloud computing, 110
culture, 26
    community, 26
    family, 26
cyber-physical system, 107

D

data-driven decision-making, 76, 82, 83
debt avenues, 92, 93
    financial technologies (FinTech), 93
    venture debt, 94

E

entrepreneur, 2
entrepreneurial attitude, 125
    characteristics, 124
    journey, 7
    resilience, 22
entrepreneurship, 2, 39, 40

F

flexible manufacturing system, 113

I

incubation, 98, 157
Industry 4.0, 80, 106, 107, 154
innovation, 13, 17, 41
Internet of Things (IoT), 108

L

lean business model, 9
lean manufacturing, 6, 7

M

multi criteria decision making (MCDM), 77
    analytic hierarchy process (AHP), 78
    analytic network process (ANP), 78
    interpretive structural modeling (ISM), 78
    PROMETHEE, 79
    structural equation modelling, 79
    TOPSIS, 79

S

seed financing, 96
social capital, 53
    bonding, 55
    bridging, 56
    linking, 56
social entrepreneur, 43, 46, 47
social entrepreneurship, 42, 43, 45
start-ups, 4, 139, 154
    environment, 91
    lean start-ups, 8
    stories, 142, 143, 144

V

venture capital, 96

W

women entrepreneurship, 174

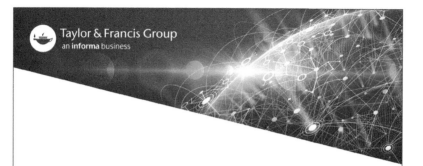